# Construction Master Workbook and Study Guide

Prepared for Calculated Industries®

and

Thomson Delmar Learning

*by*

Robert Kokernak

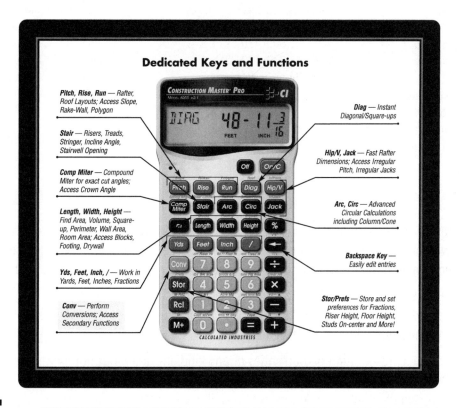

THOMSON
DELMAR LEARNING

Australia   Canada   Mexico   Singapore   Spain   United Kingdom   United States

# Construction Master® Pro Workbook and Study Guide
## Dr. Robert Kokernak, Ph.D.

**Vice President, Technology Professional Business Unit:**
Gregory L. Clayton

**Product Development Manager:**
Ed Francis

**Editorial Assistant:**
Sarah Boone

**Director of Marketing:**
Beth A. Lutz

**Marketing Specialist:**
Marissa Maiella

**Marketing Coordinator:**
Jennifer Stall

**Production Director:**
Patty Stephan

**Production Manager:**
Andrew Crouth

**Content Project Manager:**
Kara A. DiCaterino

COPYRIGHT © 2007 Calculated Industries Inc. and Thomson Delmar Learning Inc. All rights reserved. The Thomson Learning Inc. Logo is a registered trademark used herein under license. Construction Master®, Scale Master® II and Calculated Industries® are registered trademarks of Calculated Industries Inc. © 2007

Printed in the United States of America
1 2 3 4 5 XX 08 07 06

For more information contact
Thomson Delmar Learning
Executive Woods
5 Maxwell Drive, PO Box 8007,
Clifton Park, NY 12065-8007
Or find us on the World Wide Web at
www.delmarlearning.com

ALL RIGHTS RESERVED. No part of this work covered by the copyright hereon may be reproduced in any form or by any means—graphic, electronic, or mechanical, including photocopying, recording, taping, Web distribution, or information storage and retrieval systems—without the written permission of the publisher.

For permission to use material from the text or product, contact us by

Tel.  (800) 730-2214
Fax  (800) 730-2215
www.thomsonrights.com

Library of Congress Cataloging-in-Publication Data
Kokernak, Robert P.
Construction Master® Pro Workbook and Study Guide / prepared for Calculated Industries and Thomson Delmar Learning by Robert Kokernak.
p. cm.
Includes bibliographical references and index.
ISBN 1-4180-4109-2 (alk. paper)
1. Construction Master Pro (Calculator)—Handbooks, manuals, etc.
2. Building—Mathematics—Handbooks, manuals, etc. I. Title.
TH437.K58 2006
690.01′51—dc22
 2006014883

## Notice to the Reader

Publisher does not warrant or guarantee any of the products described herein or perform any independent analysis in connection with any of the product information contained herein. Publisher does not assume, and expressly disclaims, any obligation to obtain and include information other than that provided to it by the manufacturer.

The reader is expressly warned to consider and adopt all safety precautions that might be indicated by the activities herein and to avoid all potential hazards. By following the instructions contained herein, the reader willingly assumes all risks in connection with such instructions.

The publisher makes no representation or warranties of any kind, including but not limited to, the warranties of fitness for particular purpose or merchantability, nor are any such representations implied with respect to the material set forth herein, and the publisher takes no responsibility with respect to such material. The publisher shall not be liable for any special, consequential, or exemplary damages resulting, in whole or part, from the readers' use of, or reliance upon, this material.

# CONTENTS

Introduction / v
Important Notes for Owners of Previous Construction Masters / vi
Glossary of Construction Terms / vii

## PART A: SITE DEVELOPMENT — 1

**CHAPTER 1:** Scaled Distances and Areas — 3

**CHAPTER 2:** Excavation, Fill, and Grade Lines — 19

## PART B: FOOTINGS, SLABS, AND WALLS — 29

**CHAPTER 3:** Footings and Slabs — 31

**CHAPTER 4:** Walls — 41

## PART C: FRAMING — 49

**CHAPTER 5:** Walls — 51

**CHAPTER 6:** Rafters — 59

**CHAPTER 7:** Stairs — 75

## PART D: FINISHING — 81

**CHAPTER 8:** Brick — 83

**CHAPTER 9:** Roofing — 87

**CHAPTER 10:** Drywall — 91

## ANSWERS TO PROBLEMS — 95

# INTRODUCTION

Congratulations on your decision to use the *Construction Master Pro* Workbook. This text has been carefully prepared for students and educators. It is an easy-to-use course that will help you understand or teach common construction-math principals, building techniques and quickly acquire skills to utilize the *Construction Master Pro* calculator—accelerating your productivity on the job or in the classroom.

You will learn to solve routine construction challenges or undertake complex design and estimating problems. The Workbook topics are modular for ongoing reference and are developed with flexibility to work in conjunction with construction courses or as a self tutorial; from entry level to advanced studies.

The Workbook and supplemental material with this detail and depth could not have been completed without a broad base of support from industry professionals. Calculated Industries and our publishing partner, Thomson-Delmar Learning, gratefully acknowledge the author Bob Kokernak, contribution from our staffs, subject matter experts, technical reviewers plus valued consultation from the experts and educators at Journal of Light Construction (JLC Live) and the United Brotherhood of Carpenters (UBC). In addition, we sincerely appreciate guidance received from dedicated high school, vocational, apprenticeship training, and college instructors.

We thank you and wish you continuing success.

*CI Construction Education Division*

# IMPORTANT NOTES FOR OWNERS OF PREVIOUS CONSTRUCTION MASTERS

The Construction Master® Pro Workbook and Study Guide has been written with keystroke examples, illustrations and solutions using Construction Master Pro v3.1. If you are using a Construction Master Pro v3.0 it is important to understand the changes or enhancements to the v3.1 edition. The version number you are using appears next to the model number 4065 on the left side of the face of the calculator. You can also refer to your User's Guide for more information on any of the following functions.

| NEW/ENHANCED FUNCTION | DESCRIPTION |
| --- | --- |
| Accumulative Memory (M+) | — Also displays the average and the count upon repeated key presses of **M+**. |
| Arc and Circle Calculations | — Arc and Circle values can be solved by entering any two of the following values: Arc Length/Angle, Diameter/Radius, Chord Length, and Segment Rise. |
| Arched Segment Walls | — Calculate the segment walls outside the arc (including a base, if needed), or calculate the arched segment walls inside the arc, depending upon preference setting. |
| Blocks Function | — Solve for blocks based on values stored in **Length** only or in **Length** and **Height**. |
| Compound Miter | — Calculate the Miter and Bevel angles using the spring (crown) angle (this is the angle of the crown molding to the wall; previous version 3.0 used the angle of the crown molding to the ceiling). |
| Column/Cone Function | — Column/Cone Height is entered using the **Height** key instead of the **Rise** key (excludes Model #4080). |
| Footing Function | — Default footing area is 264 Square Inches (instead of 1.8 Square Feet). |
| Height | — Displays Volume, Wall Area, and Room Area only (excludes display of Area, Square-Up, and Perimeter, which are displayed in the Width function). |
| Jack Rafters | — Displays the incremental adjustment, which is the difference in rafter length from one rafter to the next. |

# GLOSSARY OF CONSTRUCTION TERMS

**Backfill**  Earth removed during excavation around foundation is replaced—based on drainage slope.
**Baluster**  Vertical member of railings on a stairway, or surrounding a deck or balcony.
**Band Joist**  (Rim Joist) A joist nailed around the edge or ends of floor or ceiling joists.
**Barge Rafter**  *See Fly Rafter.*
**Base Shoe**  (Carpet Strip) A strip of wood used to trim edge of baseboard.
**Battens**  Decorative wood strips, usually covering joints or wide boards.
**Batter Boards**  A reference line of horizontal posts used to mark corners or ends showing the level of the foundation walls and or piers—offset from the actual footings.
**Bearing Wall**  A wall that supports a specific load in addition to the weight of the wall.
**Berm**  A narrow raised curb, generally consisting of a continuous strip of formed-in-place asphalt, or lengths of precast concrete.
**Bird's Mouth**  A triangular cut in a rafter that rest on the top plate of a stud wall. Defined by a seat cut and heel cut.
**Board Foot**  A unit of lumber equal to a piece of 1 foot square and 1 inch thick or a value of 144 cubic inches.
**Board Lumber**  Yard lumber that is less than 2 inches thick but greater than 2 inches thick.
**Butt Joint**  The junction where ends of two boards meet in a square-cut joint.

**Cantilever**  A construction technique that allows for projection of horizontal members beyond the support.
**Cap Plate**  (Double Top Plate) Conventional framing member that is nailed to top plate of stud walls upper most top plate, sits atop plate.
**Cheek Angle**  A vertical side cut on the end of a hip, valley, or jack rafter that allows the rafter to fit snugly against adjoining structural members.
**Collar Beam**  (Tie Beams) A connecting beam connecting opposite pairs of rafters.
**Common Rafter**  One of perhaps many rafters all the same length that connect the eaves with the ridge board.
**Cornice**  Overhang of roof at the eaves, consisting of fascia board, soffit, and decorative moldings.
**Courses**  Layers of material usually block in a wall or shingles on a roof.
**Cripple**  Framing term for shorter vertical members usually used over opening or above and below framed windows, or in the roof.

**Dado Joint**  Joints where a dado is cut in one piece of wood to accept the end of another piece.

**Dimensional Lumber**  Lumber that exceeds board lumber in that it is 2 inches to 4.99 inches thick and 2 or more inches wide, generally used term to describe joist, rafter, stud, plank, or small timber lumber.
**Dormer**  On a roof, a 90-degree projection from the roofline, with a vertical wall for one or more windows.

**Eaves**  Overhand of roof line projecting beyond the walls.

**Fascia**  Front surface of cornice or eaves.
**Fire Stop**  Conventional wood frame blocking preventing or delaying the spread of fire and smoke usually refers to a piece of 2×4 cross blocking between studs.
**Fly Rafters**  (Rake Rafters, Barge Rafters, Rat Tails) End of rafters protruding from roof, usually supported by lookouts and sheathing.
**Footing**  Base that supports foundation, stem wall or piers, usually wider than wall.
**Frieze**  Decorative member of a cornice—ornamental band on building.
**Furring**  Narrow strips of wood attached to walls or other smooth surface to provide a fasting for other materials.

**Gable**  Vertical triangular part of a structure between slopes of a roof.
**Gable End**  End wall that has a gable.
**Gambrel**  A roof with two slope angles, steep one at the edge of building and more shallow at the center.
**Girder**  Large beam of steel, usually or a gluelam beam that supports parts of the structure above it.
**Grade**  Ground level of surrounding structure—a natural grade is original level. Finished grade is level after structure is completed and usually involves a slope for drainage.
**Gusset**  (Truss Plate) usually a flat piece of wood or metal nailed or set at a joint to give a ridged joint in wood framing or rafters.

**Hanger**  A support device for framing or any other item such as plumbing. Hangers are named after the type of support they provide, such as a joist hanger.
**Header**  (Lintel) A horizontal piece of lumber seen over a door, garage door, window, or other opening providing support for the area being spanned.
**Heel**  The cut that a rafter rests on when sitting on top plate of a wall.
**Heel Cut**  The vertical cut at the end of rafter that forms the bird's mouth with the heel.
**Hip**  The angle, convex formed by the meeting of two roof slopes—usually set at 90-degree angle.

# Glossary of Construction Terms

**Hip Rafter**  A 45-degree angle to the wall is a rafter that runs from the corner of a wall to ridge for forms the hip.

**Hip Roof**  Usually a portion of a roof that slopes up toward the center from all sides.

**Jack Rafter**  A rafter smaller in size than the common rafter, and spans the distance between the top plate and a hip or valley rafter and the ridge board.

**Jack Stud**  (Cripple Stud) A stud that does not go from top plate down to sole plate.

**Joist**  Series of 2 inch or greater parallel beams used to support floor or ceiling loads. Joist transfer load weights to other supporting members such as walls or foundation piers.

**Kerf**  A saw cut left in any material.

**Knee Wall**  The short wall joining the roof and floor in a sloping roof room usually in an attic or second story room.

**Landing**  The flat horizontal area between flights of stairs and or area at the bottom of a stairway.

**Ledger**  A board or strip of wood fastened to the side of a wall or other framing member and provides a ledge to transfer a vertical load to a horizontal member and provide support for fastening.

**Lintel**  *See* Header.

**Mansard**  A type of roof that has a steep vertical rise, close to the edge of structure and then tapers to the top with a much less severe slope.

**Mudsill**  (Sill Plate) Lowest framing member, usually mounted to the foundation.

**Mullion**  Vertical member that separates glass panes in a multipane window.

**Muntin**  Vertical and or horizontal dividers in a multipane window.

**Newel**  Main post at the foot of a stairway.

**On Center**  (O.C.) Spacing reference to denote from the center of one beam or member to the center of the next.

**Outrigger**  Rafter extensions that form cornices.

**Passing Wall**  On an exterior corner where two walls meet, the wall that extends all the way to the outside edge of the structure.

**Penny**  Historically referenced the cost of nails, currently refers to the length of nail reference in drawings indicated by a d.

**Pier**  A column of cement can be either circular or rectangular used to provide support.

**Pilaster**  A vertical rib made of concrete block or poured concrete on the inside of a long expanse of wall; used to stiffen the wall and reduce the likelihood of cracking or buckling.

**Pitch**  Slope or incline of a roof.

**Planimeter**  An instrument used to trace the outline of irregular shapes (such as wetlands) on a map to yield values of perimeter and enclosed area for these shapes.

**Plate**  Horizontal framing board either at the top or bottom of wall on which other framing members rest.

**Plate Cut**  (Seat Cut) the horizontal cut in a rafter that forms the top part of the bird's mouth.

**Plumb**  A measured state referencing exactly vertical and when referenced to the level of a floor, forms a perpendicular angle.

**Purlin**  A horizontal framing member that supports the common rafters in a roof.

**Rabbet**  Horizontal cut in a corner piece of lumber.

**Rafter**  A series of 2 inch or greater framing members supporting roof loads.

**Rail**  The horizontal covering over balustrades on a stairway.

**Rake**  The incline or slope of a gabled roof.

**Ribbon Board**  Used to support joists or beams.

**Ridge**  The horizontal line where the roof slopes meet at the highest point of the roof.

**Ridge Board**  The board running along the ridge line to which the rafters are attached.

**Ridge Cut**  The cut in the rafters that allows attachment to the ridge board.

**Rim Joist**  (Band Joist) the joist nailed across the ends of a roof or floor joist.

**Rip Sawing**  Sawing boards in the direction of grain.

**Rise**  In a stair it is the vertical distance from one tread to the next, while in roofing it is the vertical distance from the top of the double plate to the top ridge board.

**Riser**  Vertical boards between the treads of a stairway, joining treads.

**Run**  The horizontal measurement used in stairways to describe the distance between from front to back on a tread.

**Shed Roof**  A roof sloping in one direction.

**Sill Plate**  *See* Mudsill.

**Sleepers**  Embedded boards in a concrete floor providing a nailing surface for subflooring.

**Soffit**  The horizontal under portion of a cornice.

**Soleplate**  *See* Bottom Plate.

**Solid Bridging**  Used to prevent twisting of the joists, placed near the center of a span.

**Span**  The distance between structural supports, calculated based on the weight and distance of the area to be spanned and materials to be used to span the distance.

**Stringer**  (Carriage) Used in stairway construction to describe the sides of the stairway that are cut to support the treads and serve as a nailer for the risers if the stairway is enclosed.

**Stud**  A vertical framing member used in walls.

**Sub-floor**  Plywood or strand board attached to joists, finished flooring is laid over the sub-floor.

**Tail Cut**  At the bottom end of the rafter the cut made perpendicular to the ground to attached the fascia board to the rafter.

**Tail Joist**  Short joist used to frame an opening in a floor or ceiling and supported by a wall and header at the other end.

**Tenon**  A projection of one piece of wood that fits precisely into a mortice.
**Tie Beam**  *See* Collar beam.
**Tread**  The horizontal surface that you walk on in reference to a stairway.
**Truss**  A frame or jointed structure designed to act as a beam of long span.
**Truss Plate**  *See* Gusset.

**Valley**  The concave angle formed by the meeting of two sloping surfaces of a roof.
**Valley Rafter**  The rafter that runs from a wall plate at the corner of house along the roof valley, and the ridge.
**Weight Density**  Weight per unit volume of a material; usually expressed in lb/cu.ft , lb/cu.yd , or ton/cu.yd, and used to calculate total weight for some portion of a structure.

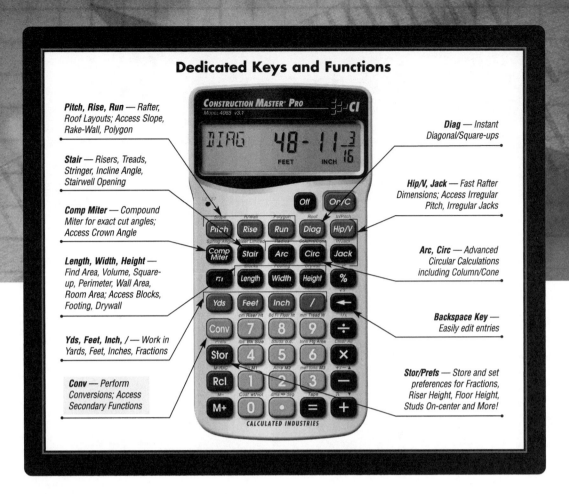

# PART A.
# SITE DEVELOPMENT

Most of the common activities associated with site development, as well as their estimated costs, are based upon linear distances, areas, or volumes determined directly from a survey map of the site. These activities may include clearing the site of trees and vegetation; excavating or filling as needed; establishing appropriate grade lines; and installing drainage, water lines, or other utilities. This section demonstrates some typical calculations encountered during the site development process.

# CHAPTER 1

# SCALED DISTANCES AND AREAS

Linear distances, used alone or to calculate areas and volumes, are generally scaled directly from a survey map or plot plan. One of the first steps taken, then, in estimating the times, costs, and amounts of material required for developing a site is to convert map distances to actual distances. Survey maps are typically drawn to scales such as 1″ = 10′, 1″ = 20′, or 1″ = 50′, with known distances given in feet, written to two decimal places (e.g., 145.73 ft). Using the appropriate portion of an *engineer's scale,* (10 divisions/inch, 20 divisions/inch, or 50 divisions/inch), actual distances may be measured directly from the map, without conversion.

Likewise, working drawings for residential and commercial structures are commonly drawn to scales such as 1/4″ = 1′, 1/8″ = 1′, or 3/32″ = 1′, with known distances given in feet and inches (e.g., 24′-6″). To measure actual distances directly from a drawing—an *architect's scale*—whose divisions match the scale of the drawing at multiples of 1/4″, 1/8″, or 3/32″, is used.

The most likely computational dilemma faced by developers, especially on-site, is trying to estimate actual distances from a site map using a ruler or tape measure having scale divisions of 1/32″ or 1/16″. To calculate actual distance, $D$, simply multiply the map distance, $d$, by the scale of the site map, $S$. In other words:

$$D\ (ft) = d\ (in.) \times S\ (ft/in.)$$

### ■ EXAMPLE 1.1
The length of pipe for a water line on a particular site map is measured as 3 - 5/8 in. If the scale of this map is 1″ = 20′, what is the actual length of pipe required?

> **Calculator Tip**
> **Entering Dimensions:**
> Enter number value, then press desired unit key ( Inch  Feet  Yds ) once for linear entry, twice for square entry, and three times for cubic entry.

> **Calculator Tip**
> Enter Fractions just as you write them: 1 / 8. See your User's Guide for changing the default Fractional Resolution.

**Solution:**

**KEYSTROKE** **DISPLAY**

First convert map distance $d = 3 - 5/8$ in. into decimal form as follows:

[3] [Inch] [5] [/] [8] [Conv] [Inch]   3.625 INCH

Re-enter as unitless value and multiply by the scale factor:

$$3.625 \times 20 = (feet\ of\ pipe)$$

[3] [.] [6] [2] [5] [×] [2] [0] [=]   72.5

### ■ EXAMPLE 1.2

What actual distance corresponds to a map distance of 7 - 9/16 in. on a map whose scale is $1'' = 50'$?

**Solution:**

[7] [Inch] [9] [/] [1] [6] [Conv] [Inch]   7.5625 INCH

Re-enter as unitless value and multiply by scale factor:

[7] [.] [5] [6] [2] [5] [×] [5] [0] [=]   378.125

The approximate linear distances scaled from a site map may now be used to estimate areas and volumes involved for various improvement activities. Calculations of areas and perimeters for rectangles and circles are easily accomplished using the built-in functions on your *Construction Master Pro* calculator.

> **Calculator Tip**
>
> **10ths, 100ths, and fraction conversions:** Multiple presses of [Inch] will convert between decimal and fractional inches; multiple presses of [Feet] will convert between decimal and feet-inch-fractions.

### ■ EXAMPLE 1.3

A mall parking lot that is 240 ft long by 165 ft wide is to be graded prior to paving. What is the area and perimeter of this lot?

**Solution:**

The multifunction [Width] key calculates area and perimeter (as well as the "square-up" diagonal distance, not needed here) using this keystroke sequence:

[2] [4] [0] [Feet] [Length]   LNTH 240 FEET 0 INCH
[1] [6] [5] [Feet] [Width]   WDTH 165 FEET 0 INCH

[Width]   AREA 39600 SQ FEET
[Width]   SQUP 291 FEET 2-15/16 INCH
[Width]   PER 810 FEET 0 INCH

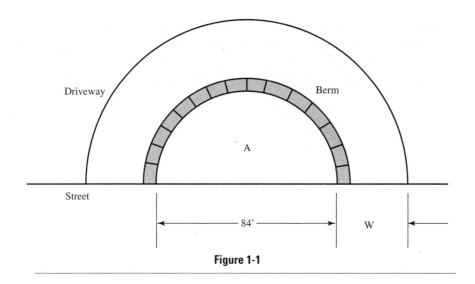

**Figure 1-1**

### ■ EXAMPLE 1.4

A semicircular driveway leading from the street to a residence and back to the street has an inside diameter of 84 ft (see **Figure 1-1**). The area between the driveway and street (**A**) is to be landscaped, and an asphalt berm (shaded) will be installed along the inside edge of the driveway. How many square feet will be landscaped; how many linear feet of berm are required?

### Solution:

**KEYSTROKE**                                                        **DISPLAY**

The area of a full circle is calculated and then divided by 2:

8 4 Feet Circ                                                        DIA 84 FEET 0 INCH
Circ ÷ 2 =                                                           2770.885 SQ FEET

Arc length (berm length) may be calculated several ways:

8 4 Feet Circ                                                        DIA 84 FEET 0 INCH
Circ Circ ÷ 2 =                                                      131 FEET 11-3/8 INCH

The berm length may also be calculated using the **Arc** function with an angle of 180 degrees:

8 4 Feet Circ                                                        DIA 84 FEET 0 INCH
1 8 0 Arc Arc                                                        131 FEET 11-3/8 INCH

Unfortunately, for site development planning and estimating, most of the areas analyzed consist of single or multiple triangles. Even irregular areas are often approximated using some combination of rectangles and triangles. The area, $A$, for all triangles may be found using the formula:

$$A = (1/2)\, bh$$

where "b" (the *base*) may be any side of the triangle, and "h" (the *height*) is the perpendicular distance from that base to the highest point on the triangle. For the triangle shown in **Figure 1-2(a),** dimension "h" may be scaled from inside or outside the triangle. **Figure 1-2(b)** shows the same triangle rotated to a new position and using a different side for "b" as well as a new dimension for "h". For the right triangle of **Figure 1-2(c),** two

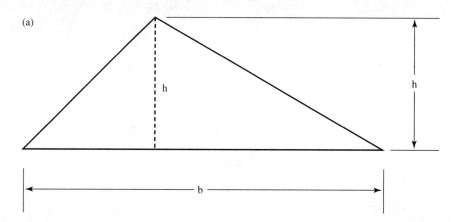

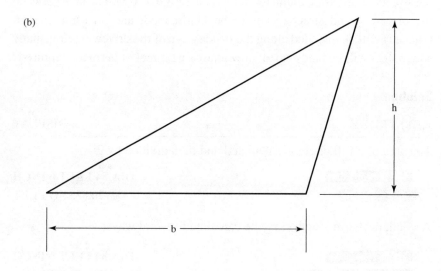

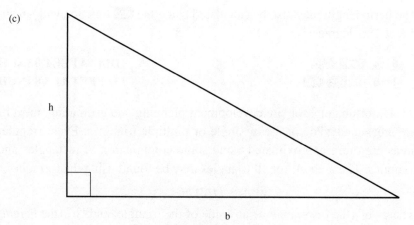

**Figure 1-2**

sides are already perpendicular, and automatically satisfy the criteria for "b" and "h."

Another method of calculating triangular areas utilizes only the three side lengths, and is particularly helpful if exact values for these lengths are already specified on the survey map or site plan. This method uses *Hero's formula*, and for triangle side lengths a, b, and c as shown in **Figure 1-3**, area A is calculated as:

$$A = \sqrt{s(s-a)(s-b)(s-c)}$$

where:

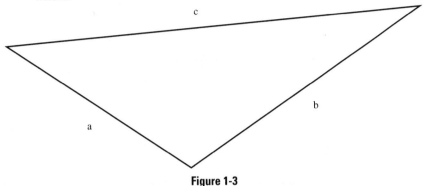

Figure 1-3

$$s = (a + b + c)/2$$

### ■ EXAMPLE 1.5

The building lot shown in **Figure 1-4(a)** is to be cleared of all vegetation. Costs for this activity are generally based on the area, in acres, for a given plot of land. Calculate the area of this lot, assuming that boundary lines a-d and b-c are parallel, and that these two lines are both perpendicular to rear lot line a-b.

Although the total area may be calculated several different ways, perhaps the simplest method is to construct an imaginary line (dashed) from point "d" perpendicular to line b-c as shown in **Figure 1-4(b)**. We now calculate the area of the rectangle, A1:

**Solution:**

| KEYSTROKE | DISPLAY |
|---|---|
| 2 0 0 • 7 6 Feet Length | LNTH 200.76 FEET |
| 2 1 9 • 4 5 Feet Width Width | AREA 44056.78 SQ FEET |

This figure may be stored in cumulative memory by pressing the **M+** key. Now clear the display by pressing **On/C** once, and calculate triangular area, A2:

| | |
|---|---|
| 1 2 3 • 6 7 Feet Length | LNTH 123.67 FEET M |
| 2 1 9 • 4 5 Feet Width Width | AREA 27139.38 SQ FEET M |

**8** Part A ■ Site Development

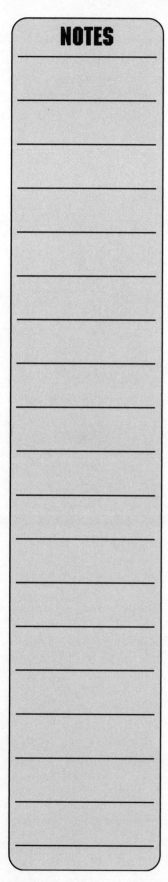

(a)

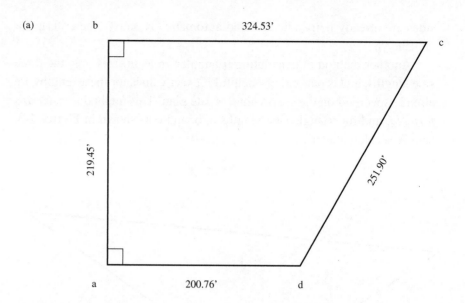

(b)

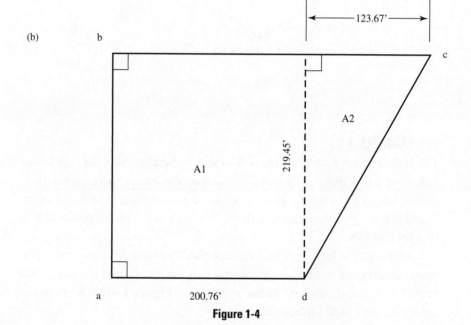

**Figure 1-4**

| KEYSTROKE | DISPLAY |
|---|---|

Convert to a triangular area by dividing this figure by 2, and then add to the rectangular area, A1, already stored in memory to display the total area, A1 + A2:

[÷] [2] [=] [M+] [Rcl] [Rcl]     M+ 57626.47 SQ FEET

This answer may be converted to acres using the [Conv] [2] sequence:

[Conv] [2] *(Acre)*     ACRE 1.322922

---

### ■ EXAMPLE 1.6

A triangular lot similar to that shown in **Figure 1-3** has these dimensions:

$$a = 51.25', b = 61.19', c = 94.39'$$

Use Hero's formula (also called Heron's formula) to find the area of this lot.

**Solution:**
First calculate "s" and store its value in semipermanent memory [M+]:

[5][1][.][2][5][+][6][1][.][1][9][+]     112.44
[9][4][.][3][9][÷][2][=][M+]     M+ 103.415 [M]

Calculate (s − a) and store this value into the permanent memory M1.

[−][5][1][.][2][5][=][Stor][1]     M-1 STORED 52.165 [M]

Clear the display and recall "s":

[On/C] [Rcl] [M+]     TTL STORED 103.415 [M]

Calculate (s − b) and store the value into the permanent memory **M2**:

[−][6][1][.][1][9][=][Stor][2]     M-2 STORED 42.225 [M]

Clear the display and recall "s":

[On/C] [Rcl] [M+]     TTL STORED 103.415 [M]

Calculate (s − c) and store the value into the permanent memory **M3**:

[−][9][4][.][3][9][=][Stor][3]     M-3 STORED 9.025 [M]

Finally, clear the display and multiply (s), (s − a), (s − b), and (s − c):

[On/C] [Rcl] [Rcl] [×] [Rcl] [1] [×] [Rcl] [2] [×] [Rcl] [3] [=]*          2055794.[11]

**KEYSTROKE**                                                                        **DISPLAY**

Taking the square root of this value yields the area of the lot, in square feet:

[Conv] [←]                                                                           1433.804

*To recall and clear the value stored in M+, press [Rcl] [Rcl] instead of [Rcl] [M+].

On a survey map, irregular areas may be found in several ways. While exact values are usually determined graphically by tracing the boundary of the area using a device called a *planimeter*, approximate values may be obtained by superimposing a grid of squares over the plot as shown in **Figure 1-5**. The number of complete squares (indicated by circles) within the perimeter are counted; fractional areas (shaded) for each square that straddles the boundary are estimated. Total area is then calculated as the sum of complete-square and fractional-square areas.

The same area may also be estimated by constructing some combination of rectangles and triangles that approximate the irregular shape as shown in **Figure 1-6**. This method is relatively accurate if partial areas outside and inside the original perimeter appear to be equal. Dimensions are scaled from the drawing, and are used to calculate areas for each of the simple geometric shapes.

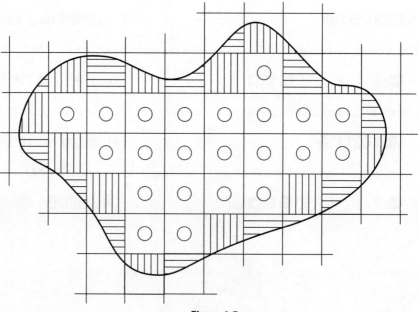

Figure 1-5

**Practical Tip**

**Scaling Plans:** Because of changes that occur during the drawing process, it is wise to be cautious when scaling directly off plans.

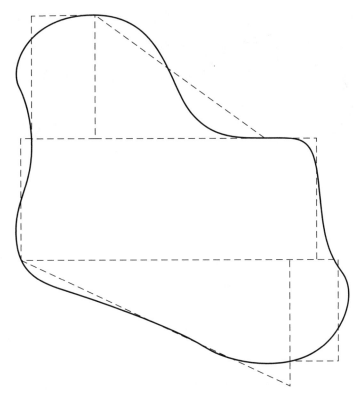

Figure 1-6

Finally, irregular areas may be divided into a series of rectangles as shown in **Figure 1-7**. If the width of each rectangle is the same, total area is equal to that width times the sum of the heights for all of the rectangles.

### ■ EXAMPLE 1.7

If scale distance "d" in **Figure 1-7** represents an actual distance of 20 ft, estimate the total area of the irregular shape shown.

### Solution:

To facilitate this area computation, use a photocopier to expand or reduce the drawing until the reference distance, "d", is one inch long. Then use an engineer's scale having 20 divisions per inch to read the width and height of each rectangle directly, to the nearest foot. The width of each rectangle is 10′, and the heights of the rectangles, working from left to right, are 18′, 50′, 64′, 61′, 53′, 50′, 54′, 57′, 40′, and 21′ respectively. Areas of these rectangles are calculated and added together as:

$(10' \times 18') + (10' \times 50') + (10' \times 64') + \ldots (10' \times 40') + (10' \times 21')$

**Practical Tip**
**Caution:** As prints and plans often go "out of scale" when printed, faxed or copied. Using a digital plan measure with a "custom scale" feature like the *Scale Master* (*www.calculated.com*) can save hours of time and prevent costly errors.

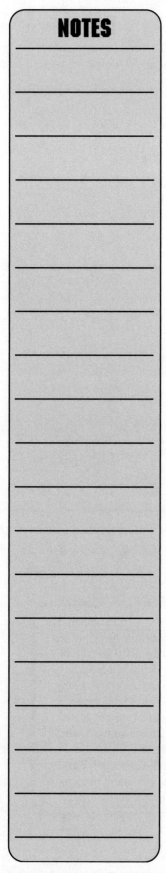

**Figure 1-7**

Algebraically, this is equivalent to:

$(10') \times [18v + 50' + 64' + 61' + 53' + 50' + 54' + 57' + 40' + 21']$

or:

**KEYSTROKE**                                                       **DISPLAY**

1 0 [Feet] × 4 6 8 [Feet] =                             **4680. SQ FEET**

## PROBLEMS FOR CHAPTER 1

**P1.1.** A survey map drawn to a scale of $1'' = 40'$ indicates a map distance of 6-3/16 in. to be trenched and backfilled for sewer pipe. How many actual linear feet does this represent?

**P1.2.** An undeveloped site is bordered on one side by wetlands. This area is to be protected by a buffer of hay bales along two straight-line sections that are 4-1/8 in. and 5-7/16 in. long on the survey map. If the map is drawn to a scale of $1'' = 30'$, what is the total number of linear feet of hay bales required for this project?

**P1.3.** The semicircular driveway in **Figure 1-1** has a width of $w = 18'$. Calculate the total area, in square feet and in square yards, to be paved.

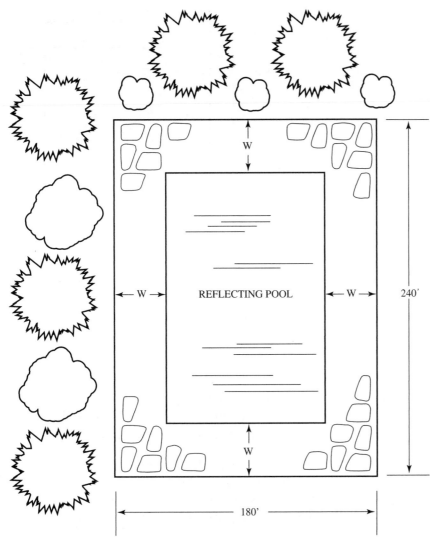

**Figure 1-8**

**P1.4.** Site development of a municipal park requires construction of a reflecting pool as shown in **Figure 1-8.** If the pool is to be surrounded by a concrete apron of uniform width $w = 24$ ft, what is the total area, in square feet and square yards, of the apron?

**P1.5.** The building complex for a new manufacturing facility has the footprint shown in **Figure 1-9.** What is the total area of these buildings?

**P1.6.** A traffic island is to be constructed in the shape of a right triangle as shown in **Figure 1-10.** The island will be paved with cobble stones and will have a granite curb around its perimeter. Find the area of this island, and the length of curbing required.

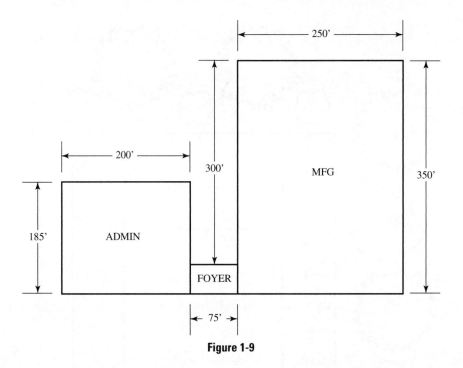

**Figure 1-9**

**P1.7.** The residential building lot shown in **Figure 1-11** is to be completely cleared of vegetation and graded. Angles A, B, C, D, and E are right angles. Find the total area of this lot.

**P1.8.** The building lot shown in **Figure 1-12** has a shape that is common where available road frontage is limited but back acreage is plentiful. If the front and rear lot lines are parallel, and the imaginary dotted lines are perpendicular to these lot lines, find the area of the lot.

**P1.9.** The side lengths of the non-right triangular area shown in **Figure 1-13** were measured on a survey map whose scale was $1'' = 20''$. Find the approximate area of this triangle.

**P1.10.** The wetlands area shown in **Figure 1-14** is to be filled in and replaced with an equal area of wetlands elsewhere on site. If the scale of this survey map is $d = 50'$, estimate the area of new wetlands that must be created.

Scaled Distances and Areas ■ Chapter 1    15

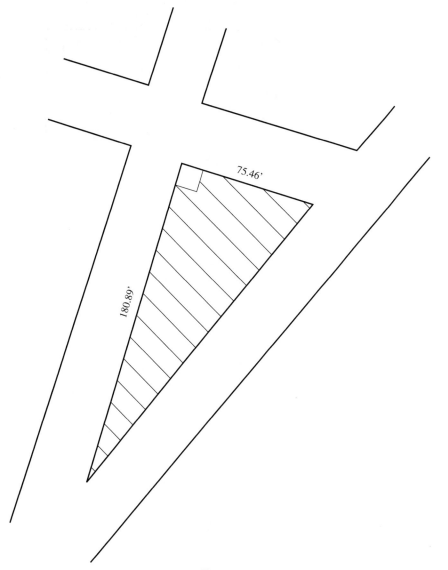

**Figure 1-10**

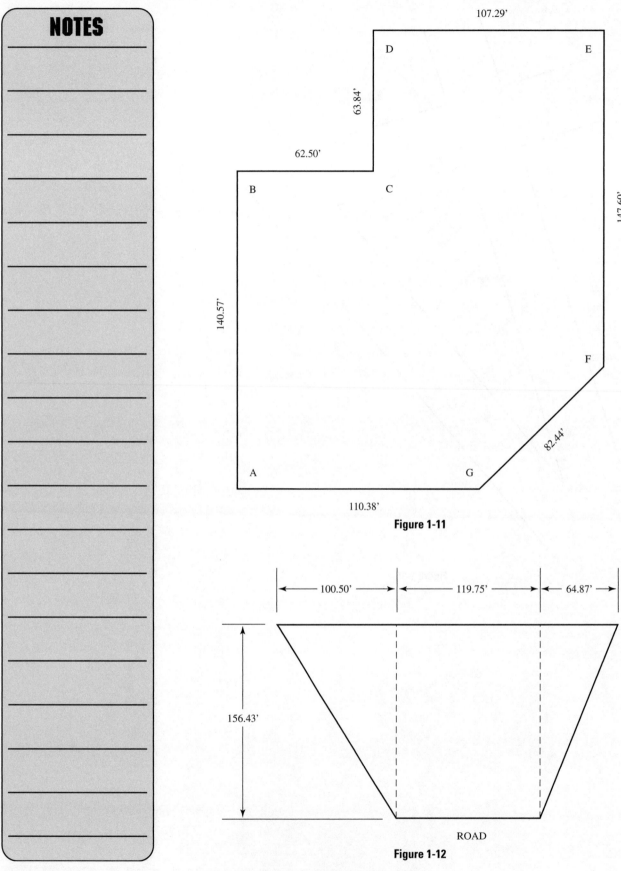

Figure 1-11

Figure 1-12

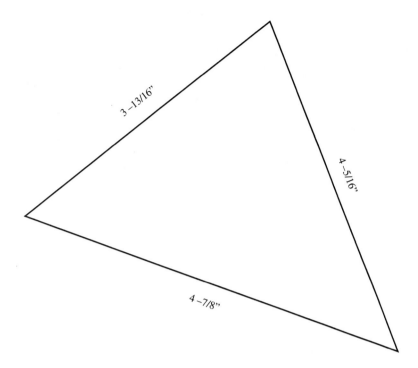

**Figure 1-13**

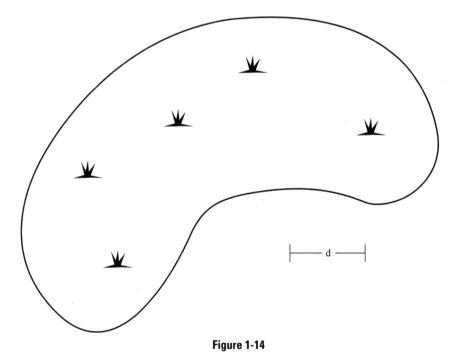

**Figure 1-14**

# CHAPTER 2

# EXCAVATION, FILL, AND GRADE LINES

Site development invariably requires excavation and removal of material, or the use of fill either brought in from outside, or "borrowed" from somewhere else on the site. Calculation of material volumes involved in these activities is a simple extension of the methods presented in Chapter 1. The general formula for calculating the volume, $V$, of any solid having a uniform cross-sectional area, $A$, is:

$$V = A \times h$$

where "h" represents the perpendicular height from the bottom surface of the solid to its highest point. Area $A$ may be a simple geometric shape as shown in **Figure 2-1 (a)**, **(b)**, and **(c)**, or an irregular shape as in **(d)** of the same figure. Also, the sides of a volume need not be perpendicular to the base; the equation above applies equally well to the solids shown in **Figure 2-2 (a)** and **(b)**. If **(a)** represents a rectangular stack of paper, for instance, its volume and height remain unchanged if the stack is pushed into the oblique configuration of **(b)**.

### ■ EXAMPLE 2.1
A rectangular area 175 ft long by 120 ft wide must be excavated to a depth of 10 ft. How many cubic yards of material must be removed? If the average tractor-trailer dump truck can carry 25 cubic yards of material, how many trips must these trucks make to this site?

### Solution:
Since the area of a rectangle is the product of length times width, the volume equation becomes:

$$V = A \times h = (L \times w) \times h$$

> **Calculator Tip**
> **Degrees and D:M:S conversions:** To convert between decimal degrees and D:M:S, press **Conv** **•**.

> **Calculator Version Note**
> **4065 v3.0:** The list of results for **Height** is based on 4065 v3.1. Users of previous models should refer to **Height** key definition in User's Guide for more details.

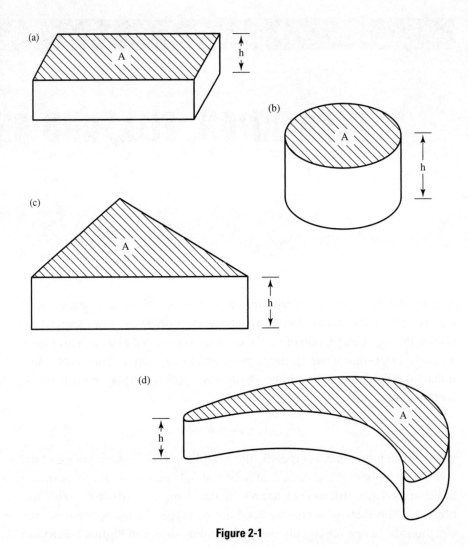

**Figure 2-1**

Performing this calculation on the *Construction Master Pro* calculator:

| KEYSTROKE | DISPLAY |
|---|---|
| 1 7 5 **Feet** **Length** | LNTH 175 FEET 0 INCH |
| 1 2 0 **Feet** **Width** | WDTH 120 FEET 0 INCH |
| 1 0 **Feet** **Height** **Height** **Conv** **Yds** | 7777.778 CU YD |

Dividing by 25 cubic yards per truckload:

| ÷ 2 5 **Yds** **Yds** **Yds** = | 311.1111 (TRUCKLOADS) |
|---|---|

### ■ EXAMPLE 2.2

As part of a site development project, the contractor plans to dig 148 ft into an *esker* (ridge of gravel) as shown in **Figure 2-3 (a)** for a distance of

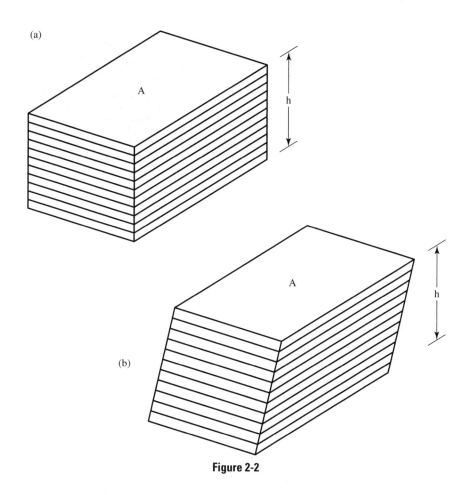

Figure 2-2

325 ft along its length. This excavated volume of material, shown in **Figure 2-3 (b)**, will be removed from the site and sold. How many cubic yards of gravel should result from this excavation? Note that, for clarity, the figure(s) are not drawn to scale.

## Solution:

**KEYSTROKE**                                                **DISPLAY**

The triangular cross-sectional area (a-b-c-a) in **Figure 2-3 (a)** is calculated using the method of Example 1.5 in Chapter 1:

1 4 8 Feet Length                                 **LNTH 148 FEET 0 INCH**
7 7 Feet Width Width ÷ 2 =                 **5698. SQ FEET**

To find the volume of excavated material, multiply this area by the length of 325 ft, and convert to cubic yards:

× 3 2 5 Feet = Conv Yds                          **68587.04 CU YD**

**Calculator Tip**

**Power User Tip:** The calculator can be set to automatically display Square or Cubic solutions in the Units of your choice; see *Setting Preferences* in your User's Guide.

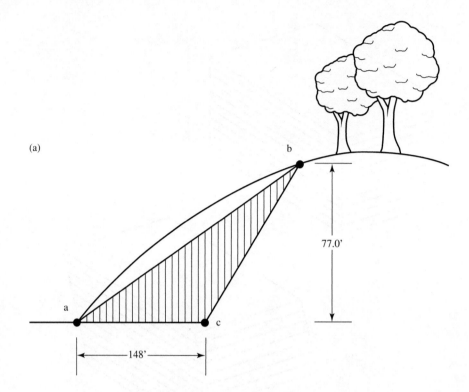

(a)

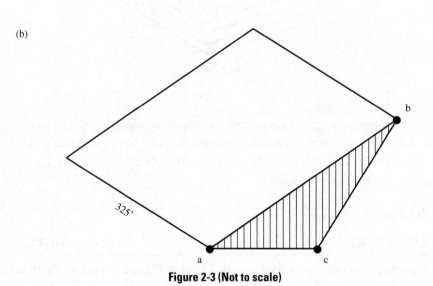

(b)

**Figure 2-3 (Not to scale)**

Although not usually associated with excavation or fill, asphalt is often part of the activities involved in development of a new site, or improvement of an existing site. While quantities of concrete, and natural materials such as sand, gravel, or loam, are sold by volume, asphalt is sold by weight. To determine the amount of asphalt needed for a particular job, this formula is used:

$$N = (0.056)\,A\,d$$

where:

$N$ = the amount of asphalt required, in tons
$(0.056)$ = a numerical constant used industry-wide to convert volumes into weights while allowing for the compressibility of a loose asphalt mixture
$A$ = the area to be paved, in square yards
$d$ = the depth of compacted asphalt required, in inches

### ■ EXAMPLE 2.3

The compacted depth of asphalt paving for the mall parking lot in Example 1.3 of Chapter 1 is to be 5-1/2 inches. How many tons of asphalt are needed?

**Solution:**

| KEYSTROKE | DISPLAY |
|---|---|

First convert the calculated area of 39,600 sq ft into square yards:

3 9 6 0 0 **Feet** **Feet** **Conv** **Yds**    4400. SQ YD

Substituting the resulting area of 4400 sq yds into the equation given above:

$$N = (0.056)(4400)(5.5) = 1355.2 \text{ (tons)}$$

**Alternate Method using Length, Width and Height keys**

By using the tons per cubic yard value commonly used for asphalt, 2 . 0 1 6 **Stor** 0*, the **Length**, **Width**, **Height** keys and weight function can be used to obtain the same results as the previous example.

Enter Length from Ex. 1.3:

2 4 0 **Feet** **Length**    LNTH 240 FEET 0 INCH

Enter Width from Ex. 1.3:

1 6 5 **Feet** **Width**    WDTH 165 FEET 0 INCH

Enter depth as Height:

5 **Inch** 1 **/** 2 **Height**    HGHT 5-1/2 INCH

Find Volume:

**Height**    VOL 672.2222 CU YD

> **Calculator Tip**
> **Power User Tip:** Alternate Method shows convenient use of built-in calculator functions.

Find weight in tons:

**Conv** **6** *(tons)*                        1355.2 TON

*If calculator does not display **Ton Per CU YD**, keep pressing the **0** key until it does. The tons per cubic yard value was obtained by using the result of solution 2.3 (**1355.2**) divided by the cubic yard value (**672.2222**) from the Alternate Method above.*

---

One of the final steps in site development work is to establish grade lines of specified steepness at various locations on the site. This may be done to provide adequate drainage, improve vehicular access, or to prepare the site for landscaping. As shown in **Figure 2-4 (a)** and **(b)**, this generally requires that existing material that is above the desired grade line be cut and removed, or that material be brought on-site to fill areas that are below grade.

The incline of a land area may be expressed in four different ways. These four defined quantities are presented here in the same order that they would appear on the display of your *Construction Master Pro* calculator.

1. *Percent Grade*—This is the most common method of expressing the grade of a land area. Referring to **Figure 2-5,** any two points, "*a*" and "*b*", on an inclined surface may be connected by a vertical line (distance b-c) called the *rise,* and a horizontal line (distance a-c) called the *run*. Percent grade for the surface is defined as the rise divided by

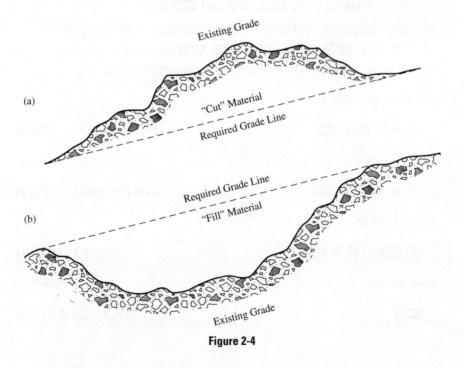

Figure 2-4

the run, expressed as a percent, where the dimensional units for both the rise and run are the same. In other words:

$$\% \text{ Grade} = \frac{\text{Rise (ft)}}{\text{Run (ft)}} \times 100\%$$

2. *Slope*—Slope is simply the percent grade expressed as a decimal, where rise and run are both measured in the same units of length:

$$\text{Slope} = \frac{\text{Rise (ft)}}{\text{Run (ft)}} = \frac{\text{Rise (in.)}}{\text{Run (in.)}}$$

3. *Pitch*—The pitch of an inclined surface is defined as the rise, in *inches*, divided by the run, in *feet*. The resulting number actually has units of "inches of rise per foot of run," but is generally expressed simply in inches:

$$\text{Pitch} = \frac{\text{Rise (in.)}}{\text{Run (ft)}}$$

4. *Pitch*—Pitch may also be expressed as the angle, *A*, in degrees, as shown in **Figure 2-5**. Using percent grade, slope, or pitch information, the *Construction Master Pro* calculator automatically computes this angle and displays it in decimal form. The angle may also be converted to degrees-minutes-seconds.

### ■ EXAMPLE 2.4

An existing grade rises 12'-8" over a horizontal distance of 45'-6". Express this incline as a percent grade, slope, pitch (in inches), pitch (in decimal degrees), and pitch (in degrees-minutes-seconds).

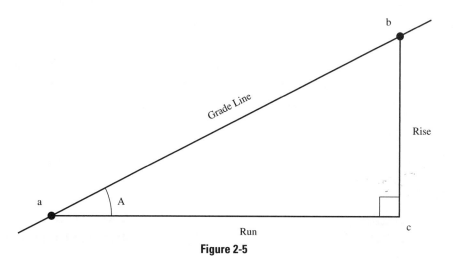

**Figure 2-5**

**Calculator Tip**

**Power User Tip:** Displayed pitch results will start with the last entered pitch format. For example, if [7] [Inch] [Pitch] is entered, any future pitch calculation will begin with the pitch in inch format, unless calculator is reset.

## Solution:

| KEYSTROKE | DISPLAY |
|---|---|

Simply enter the rise and run as follows:

| | |
|---|---|
| [1] [2] [Feet] [8] [Inch] [Rise] | RISE 12 FEET 8 INCH |
| [4] [5] [Feet] [6] [Inch] [Run] | RUN 45 FEET 6 INC |
| [Pitch] | %GRD 27.83883 |
| [Pitch] | SLP 0.278388 |
| [Pitch] | PTCH 3-5/16 INCH |
| [Pitch] | PTCH 15.56° |
| [Conv] [•] *(dms ◄ ► deg)* | DMS 15.33.24 |

> **Calculator Tip**
> **Pitch display order:** Order of answers displayed varies depending on last format of pitch entered. Continue pressing [Pitch] until desired format is displayed.

### ■ EXAMPLE 2.5

In order to promote adequate drainage away from foundations, some building codes require a minimum finish grade of 5% for a distance of at least 10 ft from the foundation. To what depth (rise) at this 10 ft distance (run) should the site be graded?

## Solution:

| KEYSTROKE | DISPLAY |
|---|---|

First, enter the known grade as:

| | |
|---|---|
| [5] [%] [Pitch] | % GRD 5. |

Next, enter the run:

| | |
|---|---|
| [1] [0] [Feet] [Run] | RUN 10 FEET 0 INCH |

Finally, press the [Rise] key

| | |
|---|---|
| [Rise] | 0 FEET 6 INCH |

### ■ EXAMPLE 2.6

For the leach field of a particular septic system, a minimum 1:3 (rise : run) grade is specified. Using a surveyor's level, the grade is measured as 17°23′. Does this satisfy the grade requirement?

## Solution:

| KEYSTROKE | DISPLAY |
|---|---|

Enter the 1:3 rise and run of the grade:

| | |
|---|---|
| [1] [Feet] [Rise] | RISE 1 FEET 0 INCH |
| [3] [Feet] [Run] | RUN 3 FEET 0 INCH |

Pressing **Pitch** which shows that the measured **grade is not adequate.**

**Pitch**                                                  18.43°

## PROBLEMS FOR CHAPTER 2

**P2.1.** The building site of Problem 1.5 in Chapter 1 is to be excavated to a depth of 12'-6". How much material must be removed?

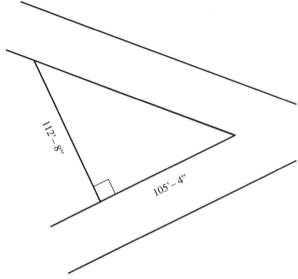

**Figure 2-6**

**P2.2.** A convenience store with self-service gasoline facilities is to be built at the intersection of two streets as shown in **Figure 2-6.** For placement of the fuel storage tanks, the triangular area (shaded in figure) is to be excavated to a depth of 14'. How much material must be removed from the site?

**P2.3.** Average depth of the irregular depression in Example 1.7 of Chapter 1 is 8'. How many cubic yards of fill are required to level the site?

**P2.4.** The driveway in Example 1.4 of Chapter 1 has a width of w = 18'. If this area is to be paved to a compacted depth of 4-1/2", how many tons of asphalt are required?

**P2.5.** Convert:

(a) 28.46° to DMS

(b) 34°12'52" to decimal degrees

**P2.6.** Find the rise of an 8% grade in a run of 510′.

**P2.7.** An existing grade rises 21.5′ over a run of 86.75′. Find the percent grade, and the pitch, in degrees.

**P2.8.** A handicapped-access ramp is limited to a 1:12 (rise:run) slope. If the ramp must reach a doorway that is 3′–4″ above ground, how far from the building must the ramp begin? If the ramp will be 10′ wide and of poured concrete, how many cubic yards of concrete will be required?

**P2.9.** To stabilize a potential building lot, the developer plans to construct a 12-ft high engineered retaining wall across the 200-ft wide lot as shown in **Figure 2-7.** As an alternative, he may add fill to change an existing grade to the "proposed" grade (dotted line) shown in the figure. For the elevation and setback dimensions given, find the percent grade of the existing and proposed grade lines. How many cubic yards of fill would be required to make this change?

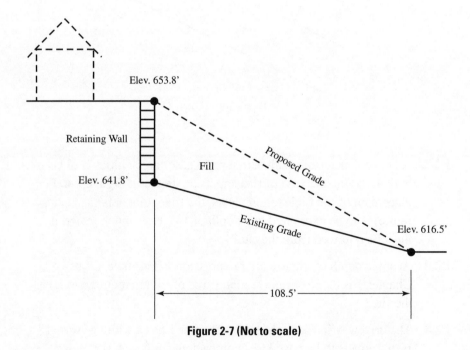

Figure 2-7 (Not to scale)

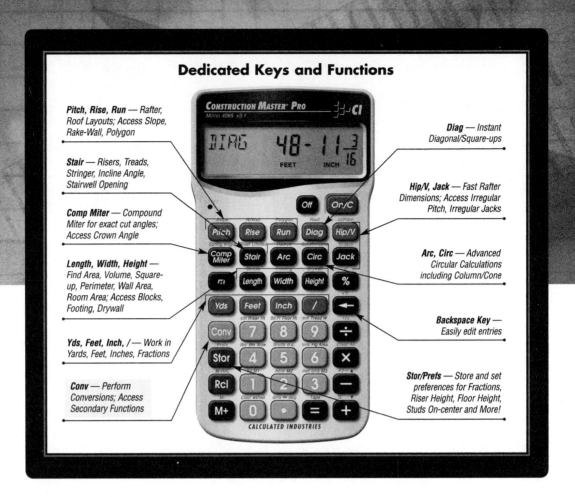

# PART B.
# FOOTINGS, SLABS, AND WALLS

Footings and foundations provide the support for virtually all residential structures, while slabs are most commonly used as floors, patios, or other flat surfaces including walkways, driveways, and parking pads. Retaining walls or foundation walls are simply thick slabs that have been stood on edge. The volumes of concrete required for such projects are easily calculated using the methods of Chapter 2, where volume was simply the product of length times cross-sectional area of a solid shape. For walls made of concrete block, the number of blocks required may be determined using programmed features of the *Construction Master Pro* calculators.

# CHAPTER 3

# FOOTINGS AND SLABS

Continuous footings such as those that support the foundation walls of a structure are essentially rectangular solids. Consider the perimeter footing shown in **Figure 3-1** (in these examples, enter d as **Width** and t as **Height**). The important dimensions for this footing are its exterior length and width, L and W; its thickness or depth, t; and its face-width, d. We may calculate either the *approximate* volume of this footing, or its *exact* volume. For the approximate method, we simply unfold the footing as shown in **Figure 3-2,** to obtain a rectangular solid whose length is the outside perimeter (2 L + 2 W), and whose cross-sectional area is (d × t). Multiplying length by cross-sectional area, the approximate volume, $V'$, of the footing becomes:

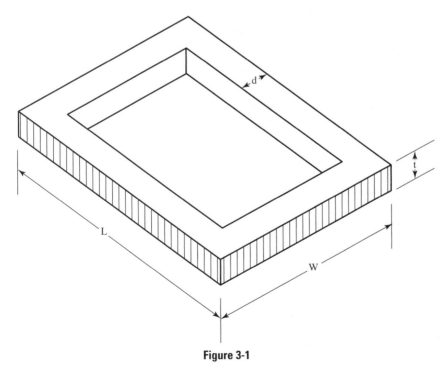

**Figure 3-1**

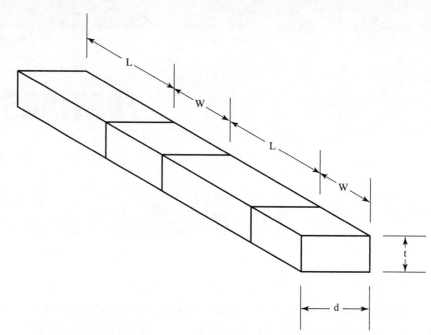

Figure 3-2

$$V' = (2L + 2W) \times (d \times t)$$

or:

$$V' = (Perimeter) \times (Cross\text{-}sectional\ Area)$$

This volume, $V'$, is higher than the exact value due to the volumes at each corner. Consider the "fit" of the side and end portions of the footing as shown in **Figure 3-3(a).** As with a mitre joint in wood, the shaded material must be removed from each section in order for the two portions to fit together. If such a footing were unfolded, its plan view would resemble **Figure 3-3(b)** rather than **Figure 3-2.**

To calculate the exact volume of a footing, it is necessary to break the total volume up into two side pieces, A and B, and two end pieces, C and D, as shown in **Figure 3-4(a).** Dimensions of each piece are shown in **Figure 3-4(b),** and the volume of each side piece and end piece is:

$$Sides\ A\ and\ B\ (each):\ (L) \times (d \times t)$$
$$Ends\ C\ and\ D\ (each):\ (W - 2d) \times (d \times t)$$

Doubling each of these quantities and combining them yields the exact volume, $V$, of the footing:

$$V = (2L + 2W - 4d) \times (d \times t)$$

or:

$$V = (Perimeter - 4d) \times (Cross\text{-}sectional\ Area)$$

The difference between $V'$ and $V$ depends upon the actual dimensions of a footing, as demonstrated in the following example.

# Footings and Slabs ■ Chapter 3

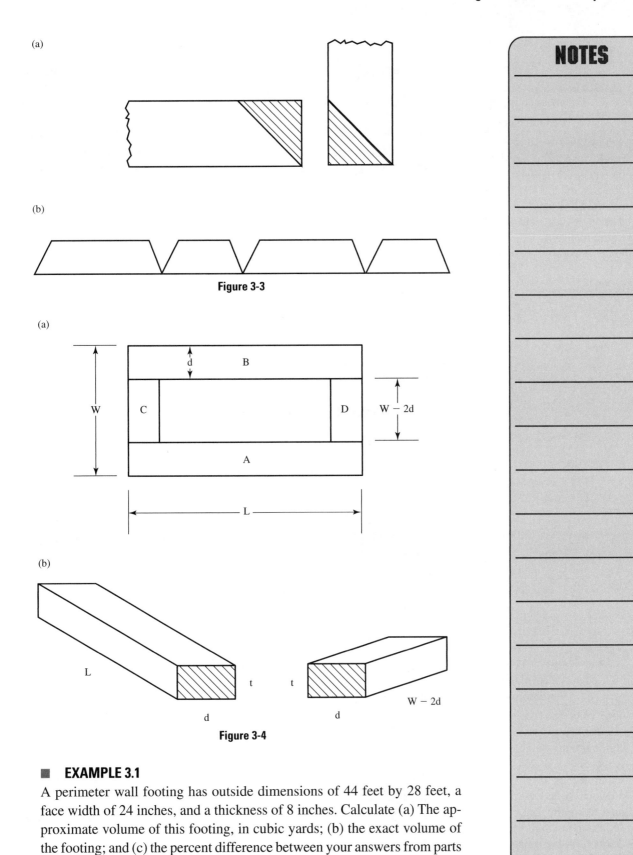

Figure 3-3

Figure 3-4

■ **EXAMPLE 3.1**

A perimeter wall footing has outside dimensions of 44 feet by 28 feet, a face width of 24 inches, and a thickness of 8 inches. Calculate (a) The approximate volume of this footing, in cubic yards; (b) the exact volume of the footing; and (c) the percent difference between your answers from parts (a) and (b).

> **Calculator Version Note**
>
> **4065 v3.0:** The list of results for [Height] is based on 4065 v3.1. Users of previous models should refer to [Height] key definition User's Guide for more details.

> **Calculator Version Note**
>
> **4065 v3.0:** The footing calculations are based on the footing cross-sectional area of 264 square inches. Users of previous models should refer to Footing key definition in User's Guide for more details.

**Solution:**

KEYSTROKE                                                                      DISPLAY

(a) First, find the perimeter of the footing:

[4] [4] [Feet] [Length]                                LNTH 44 FEET 0 INCH
[2] [8] [Feet] [Width]                                  WDTH 28 FEET 0 INCH
[Width] [Width] [Width]                               PER 144 FEET 0 INCH

Enter this value as a length:

[=] [Length]                                                 LNTH 144 FEET 0 INCH

Compute the footing volume with these additional keystrokes:

[2] [4] [Inch] [Width]                                  WDTH 24 INCH
[8] [Inch] [Height] [Height]                          VOL 7.111111 CU YD

By storing the footing cross-sectional area, [2] [4] [Inch] [×] [8] [Inch] [=] [Stor] [6], the footing function ([Conv] [6]) can be used to obtain the same results as the previous example.

**Alternate Method Using Footing Function**

KEYSTROKE                                                                      DISPLAY

The alternate solution using your calculator's footing function displays an identical answer:

[4] [4] [Feet] [Length]                                LNTH 44 FEET 0 INCH
[2] [8] [Feet] [Width]                                  WDTH 28 FEET 0 INCH
[Width] [Width] [Width]                               PER 144 FEET 0 INCH
[Conv] [Width]                                        FTG 7.111111 CU YD

(b) Begin by finding the quantity ($-4d$):

[4] [×] [2] [4] [Inch] [=] [Conv] [=] (+/−)            − 96 INCH

Add this displayed value of "− 96 INCH" to the perimeter:

[+] [1] [4] [4] [Feet] [=]                             136 FEET 0 INCH

then enter the result as a length and continue on with the volume calculation as in part (a) above:

[Length]                                                 LNTH 136 FEET 0 INCH
[2] [4] [Inch] [Width]                                  WDTH 24 INCH
[8] [Inch] [Height] [Height]                          VOL 6.716049 CU YD

| KEYSTROKE | DISPLAY |
|---|---|

Find the quantity (−4d):

[4] [×] [2] [4] [Inch] [=] [Conv] [−] [(+/−)]     − 96 INCH

Add to perimeter:

[+] [1] [4] [4] [Feet] [=]     136 FEET 0 INCH

Find footing volume:

[Conv] [Width] *(Footing)*     FTG 6.716049 CU YD

(c) The percent error is calculated as follows:

$$\frac{(7.111111 - 6.716049)}{6.716049} \times 100\% = 5.8824\%$$

The *Construction Master Pro* calculator also contains a specific keystroke sequence for approximate footing volumes that uses either a default cross-sectional area of 264 square inches (an industry standard) or the exact area for a particular footing. To utilize the default setting, simply enter the foundation perimeter as a length, and press the [Conv] [Width] keys:

| KEYSTROKE | DISPLAY |
|---|---|

[1] [4] [4] [Feet] [Length] [Conv] [Width] *(Footing)*     FTG 9.777778 CU YD

To use a stored cross-sectional area other than the 264 square inch default value, multiply the face width and thickness of the footing and store the result. For the dimensions specified in Example 3.1:

[2] [4] [Inch] [×] [8] [Inch] [=] [Stor] [6]     F - AR **STORED** 192. SQ INCH

(You should note that this value of 192 square inches will be permanently stored until you change it or reset the calculator using [Conv] [×]. To check or recall the stored setting, press [Rcl] [6].)

Now compute the approximate footing volume:

[1] [4] [4] [Feet] [Length]     LNTH 144 FEET 0 INCH
[Conv] [Width] *(Footing)*     FTG 7.111111 CU YD

This is identical to the approximate value obtained in part (a) of Example 3.1. Of course the *Construction Master Pro* calculator can be used in this manner to find exact footing values by simply modifying the above keystroke sequence to subtract (4 × d) from the 144-ft foundation perimeter.

*Point loads,* such as those transmitted by columns, are supported using discrete (non-continuous) footings. For residential use, these are generally poured in *square* forms made of 2 × 6, 2 × 8, or 2 × 10 lumber, or in *circular* forms made of commercially available waxed cardboard which has been cut to an appropriate length (footing depth). The same type of cardboard forms are also used to create concrete piers, and the method of calculating volumes is the same as that used for perimeter footings.

### ■ EXAMPLE 3.2

In many states, decks are supported on footings, which are placed below frost level. For a particular installation, 12-inch diameter concrete piers extending one foot above grade will be supported on 24-inch square footings placed 4 feet below grade as shown in **Figure 3-5.** If six of these pier/footings will be used to support the deck, how many cubic yards of concrete are required?

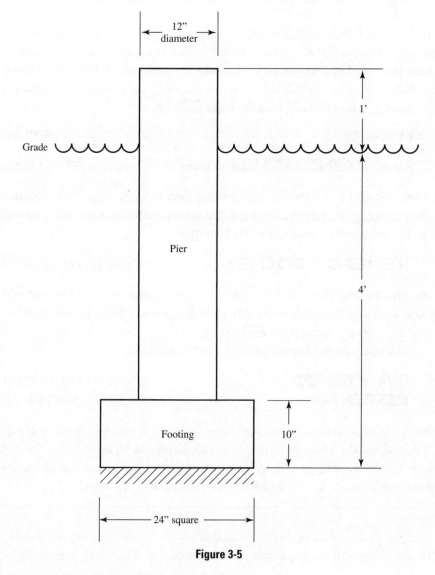

Figure 3-5

**Solution:**

| KEYSTROKE | DISPLAY |
|---|---|

Calculate the volume for one square footing:

| 2 4 Inch Length | LNTH 24 INCH |
|---|---|
| 2 4 Inch Width | WDTH 24 INCH |
| 1 0 Inch Height Height | VOL 5760. CU INCH |

Convert the displayed volume of 5760 cubic inches to cubic yards, multiply by 6 (the number of footings required), and store the result (0.740741 cubic yards) in memory:

| KEYSTROKE | DISPLAY |
|---|---|
| Conv Yds × 6 = M+ | M 0.740741 CU YD STORED |

The net height of each pier is 4′-2″. Calculate the volume of a single pier, multiply by 6 (no conversion to cubic yards required), and add to the footing volume already in memory:

| 1 2 Inch Circ | M DIA 12 INCH STORED |
|---|---|
| 4 Feet 2 Inch Height | M HGHT 4 FEET 2 INCH STORED |
| Conv Circ | M COL 0.121203 CU YD |
| × 6 + Rcl Rcl = | M+ 1.467961 CU YD |

Calculation of volumes for rectangular or square slabs may easily be accomplished using either of two different methods: multiplication of length × width × height of the slab, or by use of the multi-function Height key. Both methods are demonstrated in the following example.

### ■ EXAMPLE 3.3

The actual dimensions for a garage floor are 17′-5″ by 23′-8 1/2″. If the floor slab is to be 6″ thick, how many cubic yards of concrete are needed? Add a 5% waste allowance to your answer.

**Solution:**

| KEYSTROKE | DISPLAY |
|---|---|

Multiplying the slab dimensions for length, width, and height:

| 1 7 Feet 5 Inch | 17 FEET 5 INCH |
|---|---|
| × 2 3 Feet 8 Inch 1 / 2 | 23 FEET 8-1/2 INCH |
| × 6 Inch = | 7.646669 CU YD |

Add the specified waste allowance to produce the required volume:

| + 5 % | 8.029003 CU YD |
|---|---|

> **Calculator Version Note**
>
> **Column Height:** On older models, column height is entered using the Rise key. See Column function as described in your User's Guide.

## NOTES

**Calculator Version Note**

**4065 v3.0:** The list of results for [Height] is based on 4065 v3.1. Users of previous models should refer to [Height] key definition in User's Guide for more details.

**Calculator Tip**

**Power User Tip:** The calculator can be set to automatically display Square or Cubic solutions in the Units of your choice; see *Setting Preferences* in your User's Guide.

---

The alternate solution using your calculator's multi-function [Height] key displays an identical answer:

| | |
|---|---:|
| 1 7 [Feet] 5 [Inch] [Length] | LNTH 17 FEET 5 INCH |
| 2 3 [Feet] 8 [Inch] 1 / 2 [Width] | WDTH 23 FEET 8-1/2 INCH |
| 6 [Inch] [Height] [Height] | VOL 7.646669 CU YD |
| + 5 % | 8.029003 CU YD |

A slab may also be cast in the shape of a simple *polygon*, the most common of which are the *pentagon* (5 sides), *hexagon* (6 sides), and *octagon* (8 sides). These shapes are often used for patios, or as floors for gazebos, screen houses, bandstands, or similar structures. *Construction Master Pro* calculators allow for the quick determination of bi-sect angle, side length, perimeter, area (and therefore volume) for all such geometric shapes based upon the number of sides, N, and radius, r, of the figure.

### ■ EXAMPLE 3.4

A gazebo is to be constructed atop a hexagonal slab whose radius is 12′ as shown in **Figure 3-6**. (a) If the slab will be 8″ thick, how many cubic yards of concrete are needed? (b) If the slab shape is changed to an octagon having the same radius and thickness, how many yards are required? (c) If the slab is to be circular with the same radius and thickness, how many yards are required?

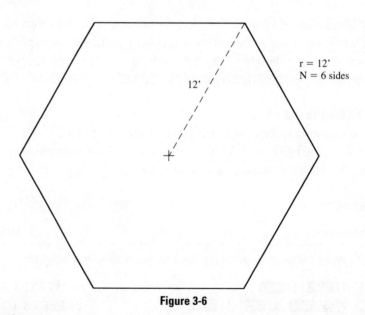

Figure 3-6

**Solution:**

(a) For any polygon, specified values of "N" and "r" are entered in this format:

( r ) [Feet] [Conv] [Arc]   *(Radius)*
( N ) [Conv] [Run]   *(Polygon)*

Successive presses of **Run** produce display values of bi-sect angle, side length, perimeter, and area respectively. Therefore, for a hexagon of 12′ radius and 8″ thickness, the volume is calculated as:

**KEYSTROKE**                                    **DISPLAY**

**1 2 Feet Conv Arc** *(Radius)*                 RAD 12 FEET 0 INCH
**6 Conv Run** *(Polygon)*                       FULL 120.00°
**Run Run Run Run**                              AREA 374.123 SQ FEET

Multiplying this value by a slab thickness of 8″ results in the required volume of concrete:

**× 8 Inch =**                                   9.237605 CU YD

(b) Similarly for an octagon:

**1 2 Feet Conv Arc** *(Radius)*                 RAD 12 FEET 0 INCH
**8 Conv Run** *(Polygon)*                       FULL 135.00°
**Run Run Run Run**                              AREA 407.2935 SQ FEET
**× 8 Inch =**                                   10.05663 CU YD

(c) For a circle of 12′ radius, the *diameter* is 24′, the volume calculation is as follows:

**2 4 Feet Circ**                                DIA 24 FEET 0 INCH
**8 Inch Height**                                HGHT 8 INCH
**Conv Circ** *(Column)*                         COL 11.17011 CU YD

> **Calculator Tip**
> **Radius:** The radius can be entered in a different unit other than feet (i.e., inches, yards, etc.)

## PROBLEMS FOR CHAPTER 3

**P3.1.** A particular residence requires a perimeter footing that is 54′- 6″ long by 26′- 8″ wide. If the footing has a face width of 28″ and a thickness of 14″, calculate the approximate number of cubic yards of concrete required. Use the *Construction Master Pro* keystroke sequence designed for this purpose, and add a 5% waste allowance to your answer.

**P3.2.** A light commercial building requires a perimeter footing that is 61′- 9″ long by 33′- 4″ wide. The footing has a face width of 30″ and a thickness of 11-1/4″. Determine: (a) The approximate volume of concrete required; (b) the exact volume of concrete required; (c) the percent difference between your answers for (a) and (b).

**P3.3.** The girder, or carrying beam, in a residential structure is supported by five steel columns, each resting on a poured concrete footing that is 18″ in diameter and 12″ thick. Find the total volume of concrete required for these footings.

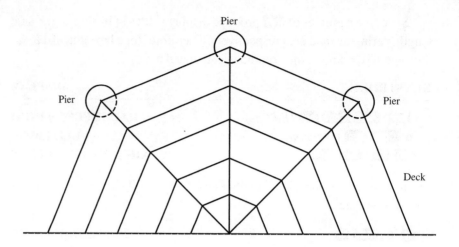

Existing Structure
**Figure 3-7**

**P3.4.** A wooden deck whose shape is half an octagon is attached to an existing structure as shown in **Figure 3-7**. The deck is supported by three piers, or columns, made of poured concrete.

(a) Each of the piers is 14″ in diameter and 72″ tall, and sits on a circular concrete footing that is 18″ in diameter and 8″ thick. Calculate the total volume of concrete required.

(b) If this deck has a radius of 16′, what is the area of the deck?

**P3.5.** A basement floor of poured concrete is 54′-4″ long, 22′-4″ wide, and 4-1/2″ thick. Find the volume of concrete required; add a 10% waste allowance to your answer.

**P3.6.** The helicopter landing pad for a hospital's med-flight aircraft is to be 60′ in diameter and 16″ thick. How many cubic yards of concrete will be required for this slab?

**P3.7.** A homeowner is considering construction of a concrete patio in the shape of a pentagon (5 sides) having a radius of 12′. (a) If the slab is to be 5-1/2″ thick, how much concrete will be needed? (b) Taxes on this residential betterment will be based on the area of the patio, but the tax assessor is unsure how to calculate this area. She estimates the value at 452.4 square feet, which is the area of a circle whose radius is 12′. What is the actual area of the patio?

# CHAPTER 4

# WALLS

Since walls of poured concrete are simply thick slabs that have been raised to a vertical position, their volumes may be calculated using the same techniques discussed in Chapter 3. There are, however, two differences between slabs and walls that must also be considered.

First, all soils have a maximum allowable contact load (called the *bearing stress*) which they can sustain over an indefinite period of time under varying weather conditions. Values of the allowable bearing stress are usually set by state or local building codes, conventional building practices in a particular geographic region, or through tests conducted by a registered professional engineer. Actual bearing stresses are created between the soil on a building site, and the bottom of a footing placed under a foundation or retaining wall. In fact, the purpose of a footing is to take the *live loads* (temporary loads such as people, furniture, wind and snow) and *dead loads* (permanent loads such as the weights of building materials) experienced by a wall, and distribute these loads over a large enough area so that the bearing stress is reduced to an acceptable value. The process is similar to putting snowshoes on your feet: Although your total weight has not been reduced, the contact area between your feet and the snow surface is increased. This reduces the force of contact (i.e., bearing stress) on the snow, and allows you to remain on the snow surface. Similarly, by using a footing whose face width is greater than the width of the wall it supports, we reduce the contact load between the wall and the ground surface. Since the load represented by a foundation wall represents a major portion of the total dead load carried by any structure, it is often necessary to calculate the *weight* of a wall so that a footing of suitable face width may be selected in order to match a structure or wall with the existing on-site soil conditions.

Also, unlike footings and slabs, which are generally made of poured concrete, foundation and retaining walls are often constructed using concrete blocks. One of the primary considerations in selecting either method of construction is *cost*. While the labor costs for installing and stripping forms used for poured concrete must be compared to the labor costs for laying up a block wall, comparison must also be made of the relative costs of materials.

In dealing with walls, then, we must not only calculate the *amount* of material required (cubic yards of poured concrete, or number of concrete blocks), but also the potential *weight* and relative material *costs* for any particular wall configuration. The following examples demonstrate these calculations using the *Construction Master Pro* calculator.

### ■ EXAMPLE 4.1

A straight wall is to be constructed of poured concrete. The wall will be 44′ long, 10′- 6″ high, and 12″ thick. (a) How many cubic yards of concrete are required? (b) If concrete weighs 150 lbs per cubic foot, find the total weight of this wall. (c) If concrete costs $97.50 per cubic yard delivered on-site, what is the total material cost for this wall?

**Solution:**

| KEYSTROKE | DISPLAY |
|---|---|

(a) Calculate the volume of the wall as follows:

| 4 4 Feet Length | LNTH 44 FEET 0 INCH |
| 1 2 Inch Width | WDTH 12 INCH |
| 1 0 Feet 6 Inch Height Height | VOL 17.11111 CU YD |

(b) For weight calculations, the given value of density or weight per unit volume of the material must first be stored in your calculator. This is accomplished using the following keystroke sequence:

1 5 0 Stor 0 *        STORED 150. LB PER CU FEET

*The units displayed with this numerical value will be one of these: "TonPer CU M" ; "kG Per CU M" ; TonPer CU YD" ; LB Per CU YD" ; or "LB Per CU FEET". If the desired units are not displayed, continue to press the (0) key until the correct value is obtained.

Now recall the volume computed in part (a):

Rcl Height Height        VOL 17.11111 CU YD

and find the total weight in pounds by pressing:

Conv 4 (lbs)        69300 LB

To convert to different weights, press the Conv key followed by 6 for tons, or by 1 for kilograms (kg). Thus:

| Conv 6 (tons) | 34.65 TON |
| Conv 1 (kg) | 31433.95 kG |

---

**NOTES**

**Calculator Version Note**

**4065 v3.0:** The list of results for Height is based on 4065 v3.1. Users of previous models should refer to Height key definition in User's Guide for more details.

**Calculator Tip**

It is often a good practice to pre-store constant values that will be used repeatedly, such as weight per volume and block area.

**Calculator Tip**

**Weight per Volume:** The calculator's default setting for weight per volume is 1.5 tons per cubic yard.

(c) The cost calculation is a straightforward multiplication:

| KEYSTROKE | DISPLAY |
|---|---|
| [Rcl] [Height] [Height] [×] [9] [7] [.] [5] [Conv] [0] (Cost) | $1668.33 |

The two most common (nominal) sizes for concrete blocks are 8″ high × 8″ wide × 16″ long and 8″ high × 12″ wide × 16″ long. Thus, each course in a concrete block wall is 8″ high, and each block in the course presents a face area of 8″ × 16″, or 128 sq. inches. This block area is a default setting in your *Construction Master Pro* calculator, and should displayed using the following key sequence:

| [Rcl] [Conv] [Length] (Blocks) | B—AR 128. SQ INCH |
|---|---|

If this 128 sq. inch value does not appear, then enter it using one of these sets of keystrokes:

| [1] [2] [8] [Inch] [Inch] [Stor] [4] | B— AR 128. SQ INCH |
|---|---|

or:

| [8] [Inch] [×] [1] [6] [Inch] [=] [Stor] [4] | B— AR 128. SQ INCH |
|---|---|

### ■ EXAMPLE 4.2

The same wall as described in Example 4.1 is to be built using concrete blocks. (a) How many concrete blocks are required? (b) If blocks with mortar weigh 85 lbs per cubic foot, what is the total weight of this wall? (c) If 8″ × 12″ × 16″ blocks cost $2.05 each, what is the total material cost for the wall?

**Solution:**

| KEYSTROKE | DISPLAY |
|---|---|

(a) First, compute the total face area of this 44′ long × 10′- 6″ high wall:

| [4] [4] [Feet] [Length] | LNTH 44 FEET 0 INCH |
|---|---|
| [1] [0] [Feet] [6] [Inch] [Width] [Width] | AREA 462. SQ FEET |

Now, find the required number of blocks:

| [=] [Conv] [Length] (Blocks) | BLKS 519.75 |
|---|---|

This rounds up to: 520 blocks

> **Calculator Tip**
>
> **Power User Tip:** 4065 v3.1 can solve for blocks based on values entered in [Length] only (for a single course), or [Length] and [Height] (for a wall area); keystrokes on older models differ. See your User's Guide for details.

(b) Using the same method described in Example 4.1(b), enter the pounds *per cubic foot* value as follows:

**KEYSTROKE**      **DISPLAY**

8 5 Stor 0      85. LB PER CU FEET

Now, enter the wall volume as computed in Example 4.1(a):

1 7 • 1 1 1 1 1 Yds Yds Yds      17.11111 CU YD

Finally, calculate the weight by pressing:

Conv 4 *(lbs)*      39270 LB

(c) Cost is calculated as the number of blocks × cost per block:

5 2 0 × 2 • 0 5 Conv 0 *(Cost)*      $1066.$^{00}$

For long expanses of concrete block walls, especially 8″ thick walls or those which have been backfilled on one side (such as most basement walls), it is common practice to *stiffen* the wall by placing 8″ × 16″ *pilasters* at 16-ft intervals along the wall. Pilasters are typically constructed as shown in **Figure 4-1,** and each requires one extra block for every course in the wall. This can substantially increase the number of blocks required for a project.

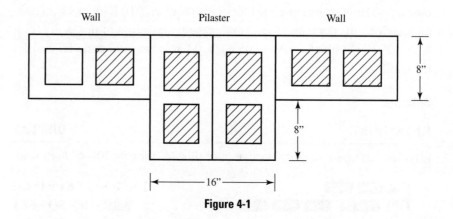

Figure 4-1

### ■ EXAMPLE 4.3

A concrete block foundation is 48′ long × 32′ wide × 9′- 4″ tall. Since the wall is 8″ thick, six pilasters are placed around the foundation at 16-ft intervals as shown in **Figure 4-2.** Find the total number of blocks required for this project, including a 3% waste allowance.

---

**Calculator Tip**

The Blocks function can also be used to determine needed quantities for counter or floor tiles, pavers, glass block, or any non-overlapping square or rectangular material. Simply set the Block Area to the size of the material desired.

Walls ■ Chapter 4   45

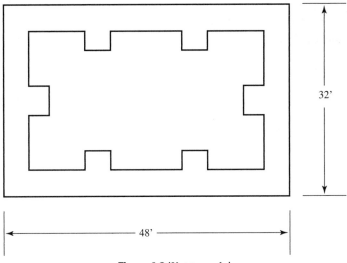

**Figure 4-2** (Not to scale)

## Solution:

**KEYSTROKE**                                                                                       **DISPLAY**

First, find the total face area of the wall as the product of *perimeter* times wall *height*.

| 4 | 8 | Feet | Length |                                                LNTH 48 FEET 0 INCH
| 3 | 2 | Feet | Width | Width | Width | Width |                  PER 160 FEET 0 INCH
| 9 | Feet | 4 | Inch | Height | Height | Height | = |          WALL 1493.333 SQ FEET

Now solve for the number of blocks needed for this wall:

---

### Calculator Version Note

**4065 v3.0:** The list of results for **Height** is based on 4065 v3.1. Users of previous models should refer to **Height** key definition User's Guide for more details.

### Calculator Tip

**Power User Tip:** If a block length is used repeatedly, store it as the block length and use the *Blocks* function; see *Block/Brick* in User's Guide for more details.

### Calculator Tip

**Clearing M+:** When using a value stored in M+ for the last time, it is often good practice to recall *and* clear it by pressing **Rcl** **Rcl**. Even though M+ has been cleared, the value remains on the display.

| KEYSTROKE | DISPLAY |
|---|---|
| [Conv] [Length] (Blocks) | BLKS 1680. |

Store this value in:

| [M+] | [M] M+ 1680. |
|---|---|

Next, determine the number of courses in the wall by dividing wall height by the 8″ height of a single course:

| [9] [Feet] [4] [Inch] [÷] [8] [Inch] [=] | [M] 14. |
|---|---|

Then each pilaster requires 14 extra blocks, so for 6 pilasters the number of blocks required is 6 × 14 = 84 blocks. Add this number to the value already in memory, and add the 3% waste allowance:

| [8] [4] [+] [Rcl] [Rcl] [+] [3] [%] | 1816.92 |
|---|---|

which rounds up to: 1817 blocks

## PROBLEMS FOR CHAPTER 4

**P4.1.** Pilasters are also used with walls made of poured concrete, especially for walls that are very high or long. These stiffeners are cast-in-place as part of the wall itself, and their sizes and locations are usually determined by an engineering analysis. A concrete wall 20′ high, 30′ long and 14″ thick has 4 pilasters located as shown in **Figure 4-3.** (a) Calculate the volume of concrete required for just the wall (no pilasters). (b) Find the total volume of concrete required for the pilasters. (c) What percentage of the total volume required for this project (wall w/pilasters) does your answer from part (b) represent?

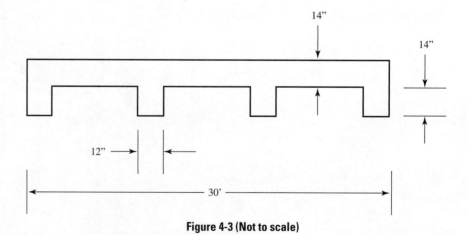

Figure 4-3 (Not to scale)

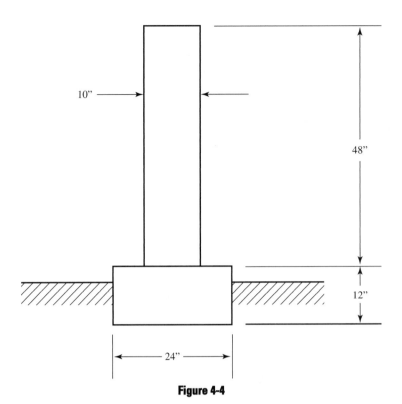

**Figure 4-4**

**P4.2.** If concrete weighs 150 lb per cubic foot, what is the total weight of the wall w/pilasters from Problem P4.1?

**P4.3.** A poured concrete retaining wall 48″ high × 10″ thick × 30′ long sits atop a footing that is 24″ wide × 12″ thick as shown in **Figure 4-4.** (a) What volume of concrete is required for the wall? (b) What volume is required for the footing? (c) If concrete costs $95 per cubic yard, what is the total cost of wall and footing?

**P4.4.** If concrete weighs 155 lb per cubic foot, what is the total weight of the wall and footing from Problem P4.3?

**P4.5.** The same footing used in **Figure 4-4** will be used under a retaining wall made from 8″ × 8″ × 16″ concrete block. The wall will still be 48″ high × 30′ long but will have 3 pilasters. (a) How many blocks are needed for this project? (b) At a cost of $1.40 per block, what will be the material cost for the wall? (c) If poured concrete costs $95 per cubic yard, what will be the material cost for the footing? (d) How does the total material cost compare with that of the wall in Problem P4.3?

**P4.6.** If concrete block w/mortar weighs 85 lb per cubic foot, and poured concrete weighs 4185 lb per cubic yard, what is the total weight of the wall and footing from Problem P4.5? How does this compare with your calculated weight from Problem P4.4?

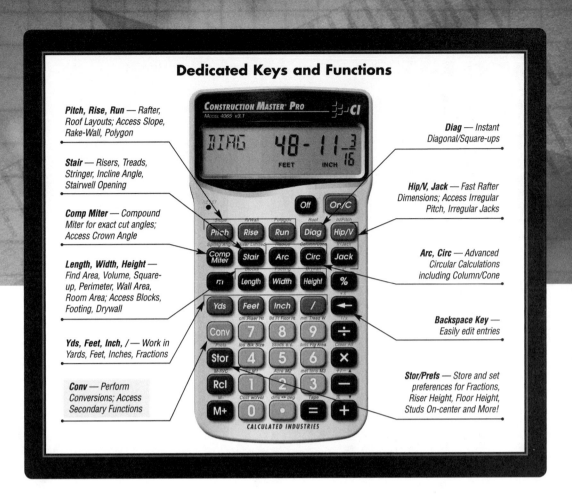

# PART C. FRAMING

The calculations involved with most framing projects are aimed at determining the following information: number of pieces of lumber required; total number of board feet; cost for the project. In addition, some structural members, such as roof rafters and stair stringers, require specialized cuts based upon trigonometric considerations. While it is not the purpose of this study guide to provide instruction in carpentry, the following chapters will describe some of the more common types of framing problems that can be solved using *Construction Master Pro* calculators.

# CHAPTER 5

# WALLS

Since most of the structural members used in the fabrication and installation of walls have squared ends, no special cuts are required, and the primary calculations are aimed at material requirements and costs. To determine the basic number of studs required to frame a wall, simply enter the length of wall, and press these keys:

**Conv** **5** *(studs)*

The calculator then displays the number of studs required, based upon an on-center (o.c.) spacing of 16 inches. The number displayed is always a whole number.

### ■ EXAMPLE 5.1

Find the number of studs required to frame a wall that is 24′ long.

**Solution:**

| KEYSTROKE | DISPLAY |
| --- | --- |

Enter the length of the wall and solve for the number of studs:

**2** **4** **Feet** **Conv** **5** *(studs)*     STUD 19

This basic number of studs is generally modified to take into account any special features in the wall. As shown in **Figure 5-1,** for example, when forming a corner, the exterior *passing wall* contains one extra stud to provide an interior nailing surface for drywall. Thus for exterior walls, one stud is added for every corner. One stud is also added for each wall segment that is not evenly divisible by 4, and for each interior partition that attaches directly to the exterior wall. Two studs are added for each window and door in a wall, this to account for conventional methods of framing the window sill with its supporting members, and the sides of the doorway opening.

> **Calculator Tip**
> **Studs On-Center:** The calculator's default setting for studs is set at 16 inches on-center.

> **Calculator Tip**
> **Stud Count:** The Studs function adds one extra stud for the end of the wall.

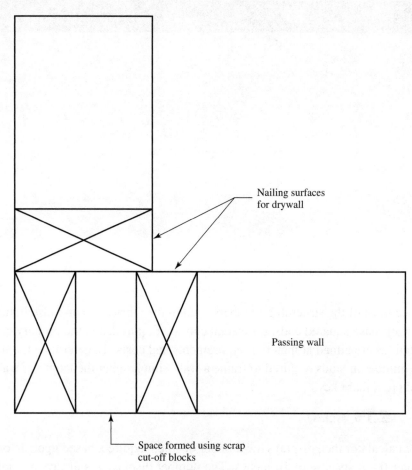

Figure 5-1

### ■ EXAMPLE 5.2

The floor plan for a typical five-room ranch house is shown in **Figure 5-2**. Estimate the number of studs required to frame the exterior walls for this structure:

**Solution:**

| **KEYSTROKE** | **DISPLAY** |

To find the basic number of studs required, use the house perimeter as the entered wall length:

| Keystroke | Display |
|---|---|
| 4 8 Feet Length | LNTH 48 FEET 0 INCH |
| 2 6 Feet 8 Inch Width | WDTH 26 FEET 8 INCH |
| Width Width Width | PER 149 FEET 4 INCH |
| Conv 5 *(studs)* | STUD 113. |

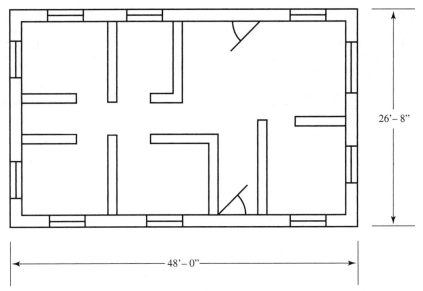

**Figure 5-2**

Now modify this number for the following features:

| Feature | Additional Studs |
|---|---|
| 4 corners | 4 |
| 2 end walls not divisible by 4 | 2 |
| 8 interior partition attachments | 8 |
| 10 windows and 2 doors | 24 |
| TOTAL | 38 |

The total number of studs needed, then, is:

$$113 + 38 = 151 \text{ studs}$$

Once the number of pieces required for a particular wall has been determined, this value can be converted to *board feet* by entering the size of the pieces, without dimensions, and pressing the keys:

**Conv** **8** *(Bd Ft)*

If the lumber cost, in dollars per thousand board feet (Mbm) is known, then total cost for wall studs is calculated by multiplying board feet by Mbm as follows:

*(Board feet)* **X** *(Mbm)* **Conv** **0** *(Cost)*

■ **EXAMPLE 5.3**

Find the total number of board feet for the studs of Example 5.2, and the total cost for these studs if each piece is 2″ × 4″ × 10′, and the lumber cost is $365 Mbm.

> **Calculator Tip**
>
> **Board Feet Entry Format:** The first two multipliers are assumed to be inches; the third multiplier is assumed to be feet. See *Board Feet* in User's Guide for more details.

**Solution:**

| KEYSTROKE | DISPLAY |
|---|---|

Find the number of board feet for each piece:

[2] [×] [4] [×] [1] [0] [Conv] [8] *(Bd Ft)*    BDFT 6.666667

For 151 studs, the total number of board feet is:

[×] [1] [5] [1] [=]    BDFT 1006.667

and the total cost is:

[×] [3] [6] [5] [Conv] [0] *(Cost)*    $367.[43]

As shown in **Figure 5-3,** all stud walls rest on a *shoe,* or *soleplate,* and are surmounted by a *double plate,* or *double cap plate.* Since these members are the same nominal size as the studs, the number of *linear feet* of lumber required for shoes and plates is simply three times the wall length. Board feet and total cost are calculated as in Example 5.3.

> **Calculator Tip**
>
> **Cost:** Unit cost, when used with board feet, is entered in the standard per thousand Board Foot measure (Mbm) format.

### ■ EXAMPLE 5.4

Find the number of board feet and total cost of the shoe and plates for the exterior walls of Example 5.3.

**Solution:**

Calculate three times the perimeter of the house:

[4] [8] [Feet] [Length]    LNTH 48 FEET 0 INCH
[2] [6] [Feet] [8] [Inch] [Width]    WDTH 26 FEET 8 INCH
[Width] [Width] [Width]    PER 149 FEET 4 INCH
[×] [3] [=]    448 FEET 0 INCH

Converting to board feet:

[2] [×] [4] [×] [4] [4] [8] [Conv] [8] *(Bd Ft)*    BDFT 298.6667

Using $365 Mbm yields the total cost of the shoe and plates becomes:

[×] [3] [6] [5] [Conv] [0] *(cost)*    $109.[01]

### PROBLEMS FOR CHAPTER 5

**P5.1.** Find the basic number of studs (no adjustments for wall features) required for walls having the following lengths:

(a) 18'- 9"     (b) 16'- 4"     (c) 22'- 0"

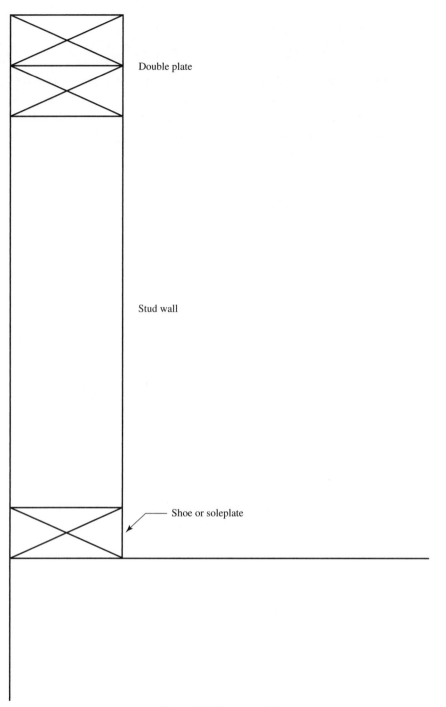

**Figure 5-3 (Not to scale)**

**P5.2.** Calculate the number of board feet in each member:

    (a) $2 \times 4 \times 8'$         (b) $2 \times 6 \times 12'$

    (c) $2 \times 8 \times 10'$       (d) $2 \times 10 \times 14'$

**P5.3.** Determine the cost for each batch of lumber:

    (a) 30 pieces of $2 \times 4 \times 10'$ @ $350 Mbm

(b) 84 pieces of 2 × 6 × 12' @ $410 Mbm

(c) 56 pieces of 2 × 12 × 8' @ $385 Mbm

**P5.4.** Find the number of board feet and the total cost to frame a single wall that is 34'- 6" long, contains two windows, and will be constructed from 2 × 6 × 10' pieces which cost $340 Mbm. Be sure to include the shoe and plates in your calculations.

**P5.5.** The sunroom addition shown in **Figure 5-4** will be framed using 2 × 4 × 8' lumber costing $415 Mbm. Calculate the total number of board feet required, as well as the total cost for framing these exterior walls.

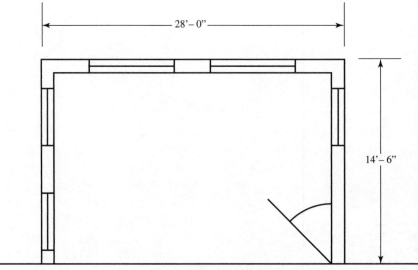

**Figure 5-4 (Not to scale)**

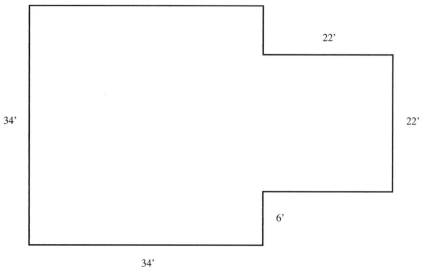

**Figure 5-5 (Not to scale)**

**P5.6.** The summer camp shown in **Figure 5-5** is to be framed using 2 × 4 × 8′ lumber costing $425 Mbm. The structure has no interior partitions, but does have 9 windows and 2 doors. Calculate the number of board feet and total cost to frame the exterior walls.

# CHAPTER 6

# RAFTERS

Placed atop framed walls, rafters support the roof of a structure. There are many types of roof styles, such as *gable, gambrel, shed, Mansard,* and *hip,* and several types of rafters, including those to be discussed here: *common, hip, valley,* and *jack* rafters.

Common rafters connect the top *plate* of a wall with the *ridge,* and are perpendicular to the ridge. Like the grade of a land area (see Chapter 2), the incline of such rafters may be specified several different ways using the quantities shown in **Figure 6-1.** The *slope* of a rafter (also called the *pitch ratio*) is a decimal value that results from dividing the rise by the run, keeping both dimensions in feet or both dimensions in inches. *Pitch* of a rafter is the rise, in inches, divided by the run, in feet, and is usually stated simply in inches. Pitch may also be expressed as angle A in the figure, that angle between the rafter and a horizontal line, and is given in degrees.

> **Warning**
> **Material Adjustments:** The rafter length calculations are the point-to-point measurements and do not account for material thickness.

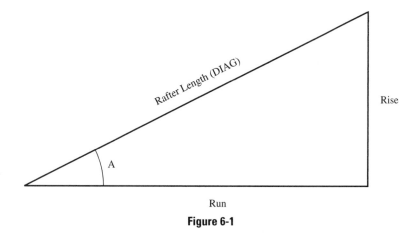

**Figure 6-1**

*Construction Master® Pro Workbook and Study Guide*

## NOTES

**Calculator Tip**

**Pitch Display Order:** Order of answers displayed varies depending on last format of pitch entered. Continue pressing [Pitch] until desired format is displayed.

**Calculator Tip**

**Power User Tip:** Displayed pitch results will start with the last entered pitch format. For example, if [7] [Inch] [Pitch] is entered, any future pitch calculation will begin with the pitch in inch format, unless calculator is reset.

### ■ EXAMPLE 6.1

A common rafter rises 5'-4" in a run of 14'-2". Find (a) the pitch of this rafter in inches, (b) its pitch, in degrees and (c) its slope.

**Solution:**

| KEYSTROKE | DISPLAY |
|---|---|

(a) (b) (c) Following the procedures outlined in Chapter 2, enter the given data as shown, and solve for the pitch:

| [5] [Feet] [4] [Inch] [Rise] | RISE 5 FEET 4 INCH |
| [1] [4] [Feet] [2] [Inch] [Run] [Pitch] | PTCH 4-1/2 INCH |
| [Pitch] | PTCH 20.63° |
| [Pitch] [Pitch] | SLP 0.376471 |

Your *Construction Master Pro* calculator can also provide the point-to-point rafter length (excluding ridge adjustment or overhang) for common rafters, as well as the angles, in degrees for making *plumb* (vertical) cuts, and *level* (horizontal) cuts as shown in **Figure 6-2.** These angles may be used for fitting rafter to ridge; cutting a *bird's mouth* to seat a rafter on the top plate; or trimming rafter tails to accept fascia and soffit boards.

### ■ EXAMPLE 6.2

For the rafter of Example 6.1, find the rafter length, and angular values for plumb and level cuts.

**Solution:**

| KEYSTROKE | DISPLAY |
|---|---|

Enter the data for rise and run as before:

| [5] [Feet] [4] [Inch] [Rise] | RISE 5 FEET 4 INCH |
| [1] [4] [Feet] [2] [Inch] [Run] | RUN 14 FEET 2 INCH |
| [Diag] | DIAG 15 FEET 1-5/8 INCH |
| [Diag] | PLMB 20.63° |
| [Diag] | LEVL 69.37° |

Using the capabilities built into your *Construction Master Pro* calculator, it is possible to determine all information related to common rafters for any specified framing conditions. Three of the situations most frequently encountered by builders are these:

1. Given the *rise* and *run,* determine the pitch and rafter length.

2. Given the *pitch* and *rise,* find the run and rafter length.

3. Given the *pitch* and *run,* calculate the rise and rafter length.

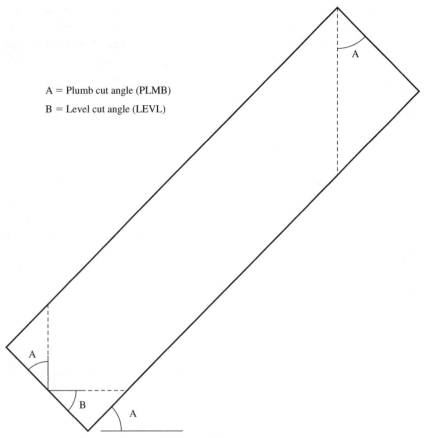

A = Plumb cut angle (PLMB)
B = Level cut angle (LEVL)

**Figure 6-2 (Not to scale)**

For each set of conditions listed above, it is a simple procedure to also find the angles required for plumb cuts and level cuts. Examples 6.1 and 6.2 demonstrated the procedure for case (1.) above in which the rise and run of a rafter are already known. The following examples demonstrate cases (2.) and (3.) respectively.

### ■ EXAMPLE 6.3

Calculate the run, length, plumb angle, and level angle for a common rafter having the following characteristics. (a) Pitch of 5-1/2″, rise of 6′-4″ (b) Pitch of 30.5°, rise of 7′-2″.

**Solution:**

**KEYSTROKE**                                                    **DISPLAY**

(a) Enter the specified conditions as follows:

[5] [Inch] [1] [/] [2] [Pitch]                             **PTCH 5-1/2 INCH**

| KEYSTROKE | DISPLAY |
|---|---|
| [6] Feet [4] Inch Rise | RISE 6 FEET 4 INCH |
| Run | RUN 13 FEET 9-13/16 INCH |
| Diag | DIAG 15 FEET 2-3/8 INCH |
| Diag | PLMB 24.62° |
| Diag | LEVL 65.38° |

(b) Similarly:

| KEYSTROKE | DISPLAY |
|---|---|
| [3] [0] [.] [5] Pitch | PTCH 30.50° |
| [7] Feet [2] Inch Rise | RISE 7 FEET 2 INCH |
| Run | RUN 12 FEET 2 INCH |
| Diag | DIAG 14 FEET 1-7/16 INCH |
| Diag | PLMB 30.50° |
| Diag | LEVL 59.50° |

### ■ EXAMPLE 6.4

Find the rise, length, plumb angle, and level angle for a common rafter having the following characteristics: (a) Pitch of 8″, run of 14′-6″ (b) Pitch of 26.6°, run of 12′-4″.

**Solution:**

| KEYSTROKE | DISPLAY |
|---|---|

(a) The solution differs only slightly from that of Example 6.3:

| KEYSTROKE | DISPLAY |
|---|---|
| [8] Inch Pitch | PTCH 8 INCH |
| [1] [4] Feet [6] Inch Run | RUN 14 FEET 6 INCH |
| Rise | RISE 9 FEET 8 INCHES |
| Diag | DIAG 17 FEET 5-1/8 INCH |
| Diag | PLMB 33.69° |
| Diag | LEVL 56.31° |

(b) Similarly:

| KEYSTROKE | DISPLAY |
|---|---|
| [2] [6] [.] [6] Pitch | PTCH 26.60° |
| [1] [2] Feet [4] Inch Run | RUN 12 FEET 4 INCH |
| Rise | RISE 6 FEET 2-1/8 INCH |
| Diag | DIAG 13 FEET 9-1/2 INCH |
| Diag | PLMB 26.60° |
| Diag | LEVL 63.40° |

*Hip rafters* run diagonally upward to the roof ridge from an *outside* corner formed by the top plates of two walls. These rafters represent the lines of intersection between two different sections of roof. *Valley rafters* also run diagonally upward, but from an *inside* corner formed by the top plates of two walls. The lines formed where two gable roofs intersect are also framed using valley rafters. **Figure 6-3** shows the plan view of an L-shaped house, with the ridges, hip, and valley rafters marked. The rafters shown as dashed lines are *hip jack rafters* (running between hip rafters and the top of a wall plate), or *valley jack rafters* (running between valley rafters and a ridge). All other rafters shown are considered to be common rafters.

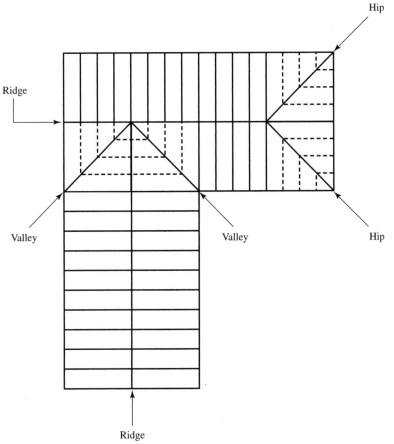

**Figure 6-3 (Not to scale)**

A *conventional* or *regular* roof contains common rafters having the same pitch on both sides of the hip or valley; *irregular* roofs have different pitches on either side of the hip or valley. In both cases, if the pitches joined by a hip rafter are the same as the pitches joined by a valley rafter, then the hip and valley rafters themselves are identical to each other in length, plumb cut, and level cut.

One additional piece of information needed for hips and valleys is the value of the *cheek angle*. This is the angle to which boards must be cut on those surfaces which mate with the ridge/common rafters, and the corner

**Warning**
The roof framing functions assume that both ridges are perpendicular to each other and that the roof is the same height throughout.

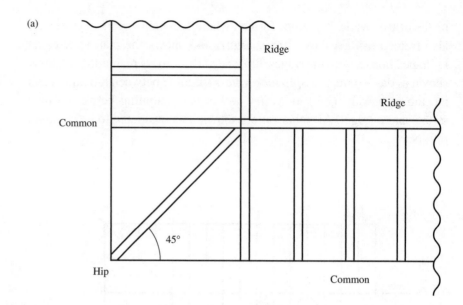

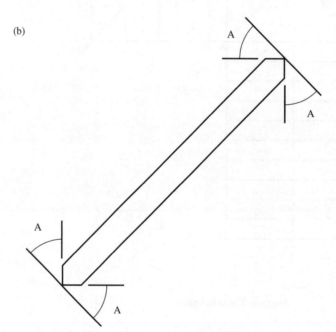

Figure 6-4 (Not to scale)

top plates. **Figure 6-4(a)** shows a plan view for the framing of a regular roof, while **Figure 6-4(b)** shows cheek angles, A, to which the ends of the hip rafter must be cut to ensure a smooth, tight fit. For regular roofs, the cheek angle is 45°. All of these lengths and angles are readily available on your *Construction Master Pro* calculator.

## EXAMPLE 6.5

The common rafters on a hip roof have a pitch of 9″, and a run of 16′-4″. Find the lengths, plumb cuts, and level cuts for the common rafters and hip rafters, as well as the cheek angle for the hip rafters.

**Solution:**

| KEYSTROKE | DISPLAY |
|---|---|

For the common rafters, use the same key sequence demonstrated in our previous examples:

| | |
|---|---|
| [9] [Inch] [Pitch] | PTCH 9 INCH |
| [1] [6] [Feet] [4] [Inch] [Run] | RUN 16 FEET 4 INCH |
| [Diag] | DIAG 20 FEET 5 INCH |
| [Diag] | PLMB 36.87° |
| [Diag] | LEVL 53.13° |

Corresponding values for the hip rafter are obtained by successively pressing the [Hip/V] key:

| | |
|---|---|
| [Hip/V] | H/V 26 FEET 1-3/4 INCH |
| [Hip/V] | PLMB 27.94° |
| [Hip/V] | LEVL 62.06° |
| [Hip/V] | CHK 1 45.00° |

In the preceding example, notice that the plumb cut for the hip rafter is significantly different from that of the common rafter. This phenomenon seems to violate all rules of common sense, and is one which perplexes many builders. (This is especially true if the common rafters are cut at an angle of 45°!) Using our definition of pitch, however, it is fairly easy to see why the hip rafter's plumb cut angle is always *less than* the plumb cut angle for its adjoining common rafters.

**Figure 6-5** shows a typical corner for a regular roof, where the common rafters have a run of Rc and both the common and hip rafters have an identical rise, Ri. Triangle A-B-C in the horizontal plane is a 45° right triangle having two sides of length Rc as shown in **Figure 6-6(a).** Side length "d" opposite the right angle, then, is greater than Rc; in fact, it may be shown using trigonometry that this side has a length of d = 1.414 × Rc. **Figure 6-6(b)** and **(c)** show the pitch triangles for the hip and common rafters respectively. The hip's pitch angle (and therefore its plumb cut angle) is Ph = Ri / 1.414 Rc, while that of the common rafter is Pc = Ri / Rc. Since the hip rafter has the same rise but a longer run than the common rafter, its pitch will always be less than that of its adjacent common rafters.

The framing members required to make the hip-plate and valley-ridge connections have varying lengths as shown for the hip rafter in **Figure 6-7.**

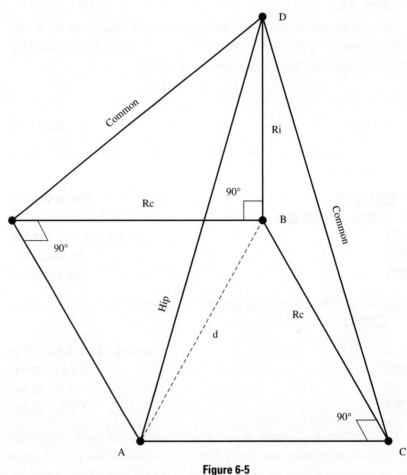

Figure 6-5

**Calculator Tip**

**Jacks On-Center:** The calculator's default setting for jacks is set to 16 inches on-center.

**Calculator Version Note**

**4065 v3.0:** Display of incremental value does not appear on this model.

The lengths of these jack rafters for 16″ center-to-center spacing are easily obtained from your calculator, as are the plumb cut, level cut, and cheek cut angles for these members. (Other spacing distances may also be used; to enter a 24″ spacing, for example, simply press [2] [4] [Inch] [Stor] [5].)

### ■ EXAMPLE 6.6

For the framing conditions of Example 6.5, find the required lengths of jack rafters at an o.c. spacing of 16″, as well as the plumb cut, level cut, and cheek cut angles for these members.

**Solution:**

Continuing on from the final displayed value (CHK1 45.00°) of Example 6.5, successively pressing the [Jack] key first displays the default o.c. spacing of 16″, and the incremental decrease in length for each jack rafter:

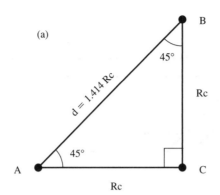
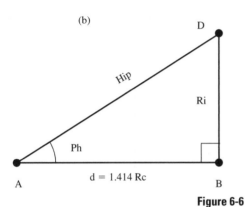
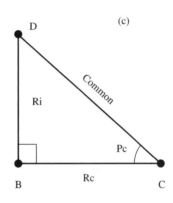

**Figure 6-6**

| KEYSTROKE | DISPLAY |
|---|---|
| Jack | STORED JKOC 16 INCH |
| Jack | INCR 1 FEET 8 INCH |

This is followed by the lengths of each jack rafter, then by the plumb cut, level cut, and cheek cut angles:

| KEYSTROKE | DISPLAY |
|---|---|
| Jack | JK 1  18 FEET 9 INCH |
| Jack | JK 2  17 FEET 1 INCH |
| Jack | JK 3  15 FEET 5 INCH |
| Jack | JK 4  13 FEET 9 INCH |
| Jack | JK 5  12 FEET 1 INCH |
| Jack | JK 6  10 FEET 5 INCH |
| Jack | JK 7  8 FEET 9 INCH |
| Jack | JK 8  7 FEET 1 INCH |
| Jack | JK 9  5 FEET 5 INCH |
| Jack | JK 10  3 FEET 9 INCH |

**Calculator Tip**

**Ascending Jacks:** Jack lengths can be set to display in ascending order; see *Setting Preferences* in User's Guide for more details.

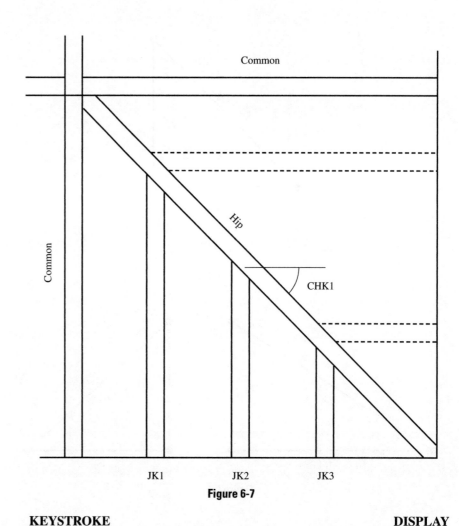

**Figure 6-7**

| KEYSTROKE | DISPLAY |
|---|---|
| Jack | JK 11  2 FEET 1 INCH |
| Jack | JK 12  0 FEET 5 INCH |
| Jack | JK 13  0 FEET 0 INCH |
| Jack | PLMB 36.87° |
| Jack | LEVL 53.13° |
| Jack | CHK 1 45.00° |

**Calculator Tip**

**Coinciding Jacks:** Regular and irregular jacks can be set to "match" or coincide; see *Setting Preferences* in User's Guide for more details.

Roofs containing two different pitches are called *irregular* or *nonstandard* roofs. Hip and valley rafters joining roof sections of different pitches each represent the diagonal of a rectangle as shown in **Figure 6-8.** For such constructions, one of the common rafters is designated as having a "regular pitch," while the other common rafter (of different pitch) is said to have an "irregular pitch." Note from the figure that the number of regular jack rafters is generally not equal to the number of irregular jack rafters, and that placement of these rafters may not "match" or coincide, even if an equal o.c. spacing is used for both sets of jack rafters. All of the required data for such hip/valley and jack rafters are readily available using your *Construction Master Pro* calculator.

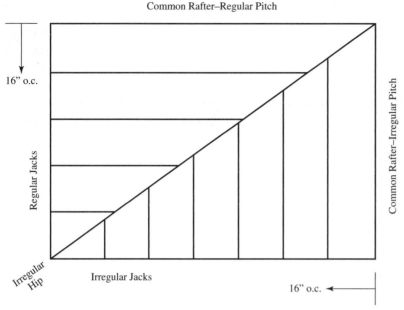

**Figure 6-8**

### ■ EXAMPLE 6.7

One section of roof having a 6-in. pitch and a run of 9′-6″ is to be joined with another roof section having an 8-in. pitch. Using the 6-in. value as the regular pitch, find the length, plumb, level, and cheek cut angles for the hip rafter and all jack rafters. Use a uniform rafter spacing of 16″ o.c..

### Solution:

| KEYSTROKE | DISPLAY |
|---|---|

First, enter the regular pitch and run, check that the calculator has a stored rafter spacing of 16″ o.c., and then enter the irregular pitch:

| | |
|---|---|
| 6 Inch Pitch | PTCH 6 INCH |
| 9 Feet 6 Inch Run | RUN 9 FEET 6 INCH |
| 1 6 Inch Stor 5 | STORED OC 16 INCH |
| 8 Inch Conv Hip/V (Ir/Pitch) | IPCH 8 INCH |

Successively pressing the Hip/V key will now display the length of your irregular hip rafter, as well as the values for plumb, level, and both cheek cut angles:

| | |
|---|---|
| Hip/V | IH/V 12 FEET 9-1/2 INCH |
| Hip/V | PLMB 21.80° |

| KEYSTROKE | DISPLAY |
|---|---|
| Hip/V | LEVL 68.20° |
| Hip/V | CHK 1 36.87° |
| Hip/V | CHK 2 53.13° |

Your calculator now provides additional information—first on the irregular jack rafters, then on the regular jack rafters—using this key sequence:

| KEYSTROKE | DISPLAY |
|---|---|
| Conv Jack | STORED IJOC 16 INCH |
| Jack | INCR 1 FEET 2-7/16 INCH |
| Jack | IJ 1  7 FEET 4-5/16 INCH |
| Jack | IJ 2  6 FEET 1-15/16 INCH |
| Jack | IJ 3  4 FT 11-1/2 INCH |
| Jack | IJ 4  3 FEET 9-1/16 INCH |
| Jack | IJ 5  2 FEET 6-5/8 INCH |
| Jack | IJ 6  1 FEET 4-1/4 INCH |
| Jack | IJ 7 0 FEET 1-13/16 INCH |
| Jack | IJ 8  0 FEET 0 INCH |
| Jack | PLMB 33.69° |
| Jack | LEVL 56.31° |
| Jack | CHK 1 36.87° |
| Jack | STORED JKOC 16 INCH |
| Jack | INCR 1  FEET 11-7/8 INCH |
| Jack | JK 1  8 FEET 7-5/8 INCH |
| Jack | JK 2  6 FEET 7-3/4 INCH |
| Jack | JK 3  4 FEET 7-7/8 INCH |
| Jack | JK 4  2 FEET 8-1/16 INCH |
| Jack | JK 5  0 FEET 8-3/16 INCH |
| Jack | JK 6 0 FEET 0 INCH |
| Jack | PLMB 26.57° |
| Jack | LEVL 63.43° |
| Jack | CHK 1 53.13° |

Notice that the jack rafters for this problem are arranged in descending order. Your *Construction Master Pro User's Guide* demonstrates the keystroke sequences required to (a) arrange the rafters in ascending order with the jacks mating at the hip/valley rafter, and (b) change the rafter spacing to something other than 16" o.c.

(a)

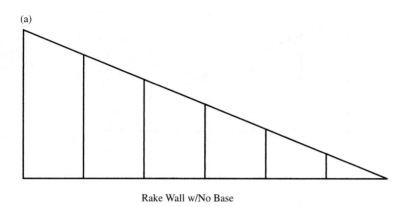

Rake Wall w/No Base

(b)

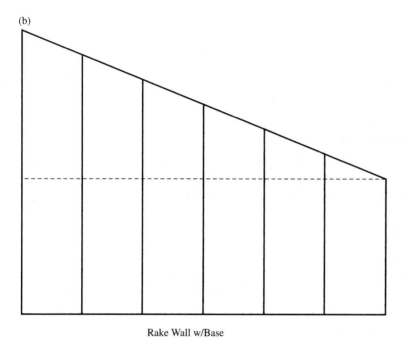

Rake Wall w/Base

**Figure 6-9**

The final structural elements to be analyzed are *rake-wall* studs as shown in **Figure 6-9.** These may be used to support the end walls of a gable roof, or the side walls of a shed roof. These members are essentially just vertical jacks, and may be installed with no base, or atop a stud wall as shown in the figure. Pitch of the roof rafters may be specified in the usual number of ways: rise and run; rise and pitch; run and pitch; or diagonal and pitch.

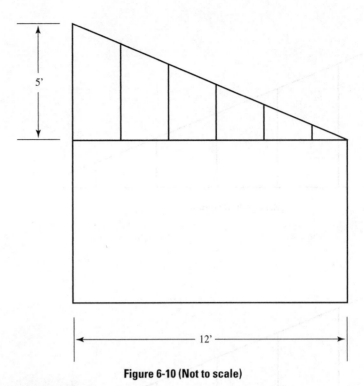

**Figure 6-10 (Not to scale)**

■ **EXAMPLE 6.8**

A rake wall having a rise of 5′ is to be added atop an old-construction stud wall that is 12′ long as shown in **Figure 6-10.** Find the lengths of rake-wall studs for a 24″ o.c. spacing, and the required cheek angle for these studs.

**Solution:**

| KEYSTROKE | DISPLAY |
|---|---|

First enter the rise, run, and o.c. spacing:

| 5 Feet Rise | RISE 5 FEET 0 INCH |
| 1 2 Feet Run | RUN 12 FEET 0 INCH |
| 2 4 Inch Stor 5 | STORED OC 24 INCH |

Accessing the rake wall function Conv Rise (*R/Wall*) and then successively pressing the Rise key will display the required stud lengths, as well as the rake-wall angle, which is also the cheek angle to which each stud should be cut:

| Conv Rise (*R/Wall*) | STORED RWOC 24 INCH |
| Rise | RW 1  4 FEET 2 INCH |
| Rise | RW 2  3 FEET 4 INCH |
| Rise | RW 3  2 FEET 6 INCH |
| Rise | RW 4  1 FEET 8 INCH |
| Rise | RW 5  0 FEET 10 INCH |
| Rise | BASE 0 FEET 0 INCH |
| Rise | RW 22.62° |

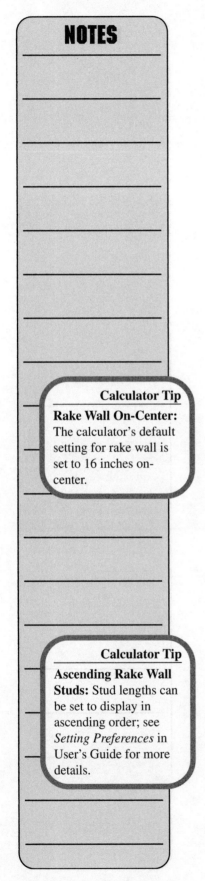

**Calculator Tip**

**Rake Wall On-Center:** The calculator's default setting for rake wall is set to 16 inches on-center.

**Calculator Tip**

**Ascending Rake Wall Studs:** Stud lengths can be set to display in ascending order; see *Setting Preferences* in User's Guide for more details.

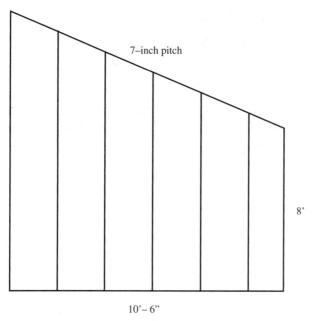

**Figure 6-11 (Not to scale)**

## ■ EXAMPLE 6.9

A rake wall having an 8′ base, 10′-6″ run, and 7-in. pitch is shown in **Figure 6-11.** Find the stud lengths for 16″ o.c., and the rake-wall angle.

**Solution:**

KEYSTROKE                                                                 DISPLAY

Enter the pitch, run, o.c. spacing, and height of base as follows:

| Keystroke | Display |
|---|---|
| 7 Inch Pitch | PTCH 7 INCH |
| 1 0 Feet 6 Inch Run | RUN 10 FEET 6 INCH |
| 1 6 Inch Stor 5 | STORED OC 16 INCH |
| 8 Feet Conv Rise (R/Wall) | BASE 8 FEET 0 INCH |
| Rise | STORED RWOC 16 INCH |
| Rise | RW 1  13 FEET 4-3/16 INCH |
| Rise | RW 2  12 FEET 6-13/16 INCH |
| Rise | RW 3  11 FEET 9-1/2 INCH |
| Rise | RW 4  11 FEET 0-3/16 INCH |
| Rise | RW 5  10 FEET 2-13/16 INCH |
| Rise | RW 6  9 FEET 5-1/2 INCH |
| Rise | RW 7  8 FEET 8-3/16 INCH |
| Rise | BASE  8 FEET 0 INCH |
| Rise | RW 30.26° |

---

**Calculator Tip**

**Rake Wall Base:** To include a rake wall base, the length must *always* be entered immediately before pressing **Conv Rise**.

## PROBLEMS FOR CHAPTER 6

**P6.1.** The common rafters on a particular structure have a 5-1/2" pitch and a run of 13'-4". Find the rise, length, plumb and level cut angles for these rafters.

**P6.2.** For common rafters having a rise of 8'-5" and a run of 15'-3", determine the pitch (in inches and degrees), length, plumb and level cut angles.

**P6.3.** Common rafters on a shed roof have a 4" pitch and a rise of 3'-8". Calculate the run, length, plumb and level cut angles for these rafters.

**P6.4.** A conventional roof has two sections containing common rafters with a 6" pitch and 15'-9" run. For these common rafters, find the pitch (in degrees), length, plumb and level cut angles. Also determine the length; plumb, level, and cheek cut angles for hip and valley rafters used to join the two roof sections.

**P6.5.** For the roof of problem P6.4, calculate all hip jack lengths for a spacing of 16" o.c. . Repeat for 24" o.c. .

**P6.6.** A conventional roof has a 12" pitch and an 8' run. Calculate the pitch (in inches), length, plumb and level cut angles for all hip/valley rafters used on this roof.

**P6.7.** For the roof of problem P6.6, find the lengths of all jacks at a spacing of 24" o.c. .

**P6.8.** An irregular roof contains one section whose common rafters have a 5" pitch and a run of 11'-8", and another section having a pitch of 6-1/2". Find the length, plumb, level and cheek cut angles for all regular and irregular jacks to be used on this roof if their spacing is 24" o.c. throughout.

**P6.9.** A rake wall with no base has a 4'-6" rise and a 14'-0" run. Find the lengths of all rake-wall studs at 24" o.c., as well as the cheek angle to be cut at the top of each stud.

**P6.10.** A rake wall with an 8'-4" base has a 5-1/2" pitch and a run of 9'-10". Determine the rake-wall stud lengths at 16" o.c., and the cheek cut angle for these studs.

# CHAPTER 7

# STAIRS

The three structural elements of a straight stairway are shown in **Figure 7-1.** *Treads* carry foot traffic up and down the stairs, and, in turn, are supported by sidepieces called *stringers*. The rectangular openings between treads and stringers are covered by vertical curtain pieces known as *risers*. *Tread width, riser height,* and *stair inclination angle* are all defined as shown in the figure. Other important design aspects to be considered in construction of a straight stairway are shown in **Figure 7-2.** Total *rise* is the vertical distance floor-to-floor or landing-to-landing with the finish flooring installed. Total *run* is the horizontal end-to-end distance of the stringer. Size of the *stairwell opening* is determined by both the *headroom* required and the total *floor thickness*.

> **Calculator Tip**
>
> **Riser Limited:** Use the *Riser Limited* function if the stored riser height should not be exceeded due to code limitations. See User's Guide for details.

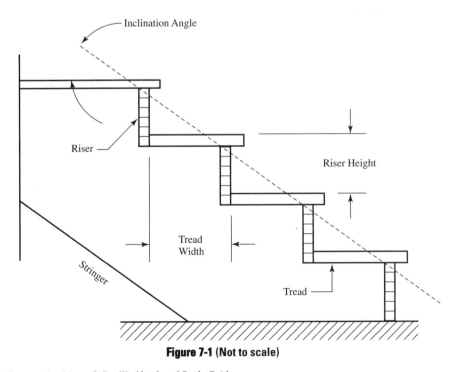

**Figure 7-1 (Not to scale)**

Construction Master® Pro Workbook and Study Guide

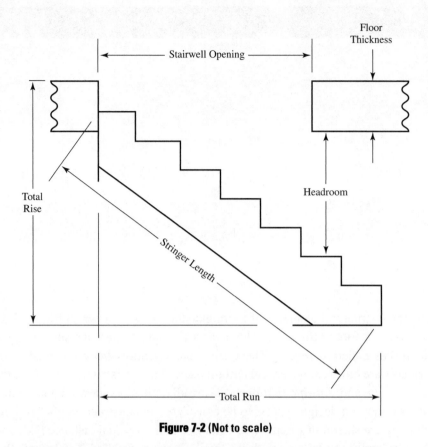

**Figure 7-2 (Not to scale)**

There are several generally accepted rules for stairway design, most of which are reflected in local building codes.

1. All steps on a stairway must have the same height and the same tread width.

2. The *sum* of riser height and tread width most commonly found on stairways is 17-1/2″. For this reason, the default values for riser height and tread width that are preset on your *Construction Master Pro* calculator are 7-1/2″ and 10″ respectively.

3. The maximum recommended riser height is 8-1/4″, and minimum recommended tread width is 9″.

4. Minimum recommended headroom is 6′-8″, which is the default value that is preset on your calculator.

5. The average value used for floor thickness is 10″, and so this is the preset default value on your calculator as well.

The challenge to good stair design lies in satisfying the above criteria as closely as possible within the vertical and horizontal space available. In the past, this task was usually accomplished using construction handbooks that contained tabulated values for various combinations of rise, run, tread width, and riser height. Today, that same information is available electronically using the *Construction Master Pro* calculator. In addition, this calculator allows the operator to specify any desired values for the design parameters, and not be restricted to those default values described above.

Two of the most common types of stair construction projects involve the following design conditions: (1) given the total rise, with no restrictions on total run; and (2) given both the total rise and total run. The following examples demonstrate the general methods used for such projects.

## ■ EXAMPLE 7.1

A straight stairway is to be built that connects the first and second floors of a new residence under construction. The floor-to-floor distance is 8'-6", and because the stairway will be adjacent to an outside wall, there is no restriction on the length of run. Using the preset values in your calculator (riser height of 7-1/2", tread width of 10", floor thickness of 10", and headroom of 6'-8") determine the following values:

(a) Actual riser height, number of risers, and overage/underage. (Note that because the number of risers may not always divide evenly into a given rise, the sum of riser heights may be over or under the specified rise by a fraction of an inch.)

(b) Actual tread width, number of treads, and overage/underage.

(c) Length of stairwell opening.

(d) Stringer length.

(e) Angle of inclination.

(f) Total run.

**Solution:**

| KEYSTROKE | DISPLAY |
|---|---|

First, input the specified rise:

| 8 Feet 6 Inch Rise | RISE 8 FEET 6 INCH |

Next, check the default values for riser height, tread width, floor thickness, and headroom:

| Rcl 7 | STORED R-HT 7-1/2 INCH |
| Rcl 9 | STORED T-WD 10 INCH |
| Rcl 8 | STORED FLOR 10 INCH |
| Conv Stor *(Prefs)* Stor Stor Stor | HDRM 6 FEET 8 INCH |
| Stair | R-HT 7-5/16 INCH |
| Stair | RSRS 14 |
| Stair | R+/− 0-3/8 INCH |

> **Practical Tip**
>
> **Riser/Tread +/− Values:** This overage or underage may be added to or subtracted from a single riser/tread, or spread across several.

| KEYSTROKE | DISPLAY |
|---|---|
| Stair | STORED T-WD 10 INCH |
| Stair | TRDS 13. |
| Stair | T+/− 0 INCH |
| Stair | OPEN 10 FEET 3-1/16 INCH |
| Stair | STRG 13 FEET 5-1/16 INCH |
| Stair | RUN 10 FEET 10 INCH |
| Stair | STORED RISE 8 FEET 6 INCH |

These displayed values tell us that there is one more riser than tread (always true), and that the sum of the riser heights is 3/8″ more than the specified rise of 8′-6″ (14 risers × 7-5/16″ per riser = 102-3/8″ = 8′-6 3/8″). Because there is no restriction on total run, the sum of tread widths can be any value and still have a zero overage/underage.

To try different solutions for the same design problem, it is only necessary to change the stair parameter settings in your calculator. You should note, however, that any newly installed values become permanent default settings until new values are installed.

### ■ EXAMPLE 7.2

Repeat Example 7.1 using these parameters: riser height of 7″; tread width of 10-1/2″; floor thickness of 11″; and headroom of 7′-0″.

**Solution:**

| KEYSTROKE | DISPLAY |
|---|---|

Enter the new stair parameters as follows:

| KEYSTROKE | DISPLAY |
|---|---|
| 7 Inch Stor 7 | STORED R-HT 7 INCH |
| 1 0 Inch 1 / 2 Stor 9 | STORED T-WD 10-1/2 INCH |
| 1 1 Inch Stor 8 | STORED FLOR 11 INCH |
| Conv Stor Stor Stor Stor + + + + | HDRM 7 FEET 0 INCH |

(Note that for headroom settings, each time the ✚ key is pressed, the headroom value already in storage increases by 1″; pressing the ✖ key decreases the existing value by 1″.)

| Stair | R-HT 6-13/16 INCH |
|---|---|
| Stair | RSRS 15. |
| Stair | R+/− 0-3/16 INCH |
| Stair | STORED T-WD 10−1/2 INCH |

| KEYSTROKE | DISPLAY |
|---|---|
| Stair | TRDS 14. |
| Stair | T+/− 0 INCH |
| Stair | OPEN 12 FEET 2-7/16 INCH |
| Stair | STRG 14 FEET 7-1/4 INCH |
| Stair | INCL 32.98° |
| Stair | RUN 12 FEET 3 INCH |
| Stair | STORED RISE 8 FEET 6 INCH |

### ■ EXAMPLE 7.3

A straight stairway is to have a total rise of 10′-4″ and a total run of 12′-6″. Using the standard values of 7-1/2″ for riser height, 10″ for tread width and floor thickness, and 6′-8″ for headroom, calculate all important dimensions required for construction of this stairway.

**Solution:**

| KEYSTROKE | DISPLAY |
|---|---|

Making sure that the proper values of riser height, tread width, floor thickness, and headroom are stored in your calculator, enter the rise and run:

| KEYSTROKE | DISPLAY |
|---|---|
| 1 0 Feet 4 Inch Rise | RISE 10 FEET 4 INCH |
| 1 2 Feet 6 Inch Run | RUN 12 FEET 6 INCH |
| Stair | R-HT 7-5/16 INCH |
| Stair | RSRS 17. |
| Stair | R+/− 0-5/16 INCH |
| Stair | T-WD ⚠ 9-3/8 INCH |
| Stair | TRDS 16. |
| Stair | T+/− 0 INCH |
| Stair | OPEN 9 FEET 7-3/8 INCH |
| Stair | STRG 15 FEET 10-1/4 INCH |
| Stair | INCL 37.95° |
| Stair | STORED RUN 12 FEET 6 INCH |
| Stair | STORED RISE 10 FEET 4 INCH |

Note that the symbol ⚠ appears when the calculated tread width is less than the stored tread width. This same symbol also appears on the display if the calculated riser height exceeds the stored riser height.

## PROBLEMS FOR CHAPTER 7

**P7.1.** A new-construction house has floor-to-floor rises of 9′-4″, and is designed for standard values of 7-1/2″ riser height, 10″ tread width and floor thickness, and 6′-8″ headroom. Calculate the necessary stair dimensions if there are no restrictions on total run of the stairs.

**P7.2.** The house of P7.1 is redesigned by the architect because his client is taller and heavier than average. For this reason, the floors will be 12″ thick, required headroom will be 7′-2″, riser height will be reduced to 7″, and tread width increased to 10-1/2″. Calculate the new stair dimensions if there are still no restrictions on total stairway run.

**P7.3.** The total run of a stairway is limited to 16′-0″. For standard values of riser height, tread width, floor thickness, and headroom (see P7.1), determine all required stair dimensions.

**P7.4.** Using the standard values of riser height, tread width, floor thickness, and headroom (see P7.1), find all required parameters of a stairway having a rise of 7′-6″ and a run of 10′-9″.

**P7.5.** In a standard ranch-style house, stairs to the basement generally descend from a central corridor in a direction perpendicular to one of the structure's side foundation walls. For this reason, such stairways are often quite steep. Using the maximum recommended riser height of 8-1/4″ and the minimum recommended tread width of 9″, find the important dimensions of a stairway having a total rise of 9′-4″. Assume a 10″ floor thickness and 6′-8″ of headroom.

**P7.6.** For the elevated deck around an above-ground swimming pool, total rise from a concrete pad to the deck surface is 7′-6″. For a riser height of 7-1/2″ and a tread width of 11″, find the required stairway dimensions. (Although none are needed, assume values of 10″ for the floor thickness and 6′-8″ of headroom.)

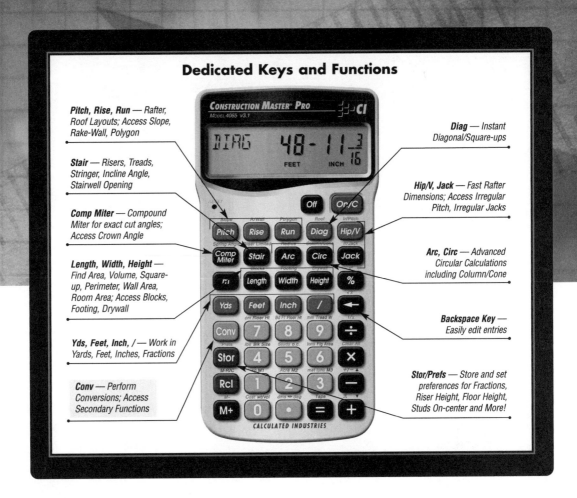

# PART D.
# FINISHING

Three common aspects of the finishing process involve brickwork, roofing, and interior drywall. As with the concrete blocks discussed in Chapter 4, the amount/number, estimated cost, and, in some instances, total weight of finishing materials may be needed for a particular project. The methods of calculation are similar to those for concrete block, and all are based on the total area to be finished.

# CHAPTER 8

# BRICK

The nominal size of a standard brick is 8″ long × 4″ wide × 2″ high. Including mortar, the "face" area covered by a single brick is generally taken as 21 square inches. To calculate the number of bricks required to face any given surface, then, this 21 square inch area is compared to the total area of that surface.

### ■ EXAMPLE 8.1

A straight wall 20′-6″ long and 8′-4″ high is to be faced with brick. How many bricks are required? At a cost of $0.45 per brick, what is the total material cost (excluding mortar) for this wall?

**Solution:**

| KEYSTROKE | DISPLAY |
|---|---|

Using the same keystroke sequence as for concrete block, enter the 21 square inch coverage for a single brick:

2 1 *Inch* *Inch* **Stor** 4        **STORED** B—AR 21. SQ INCH

Next enter the length and height of the wall:

2 0 **Feet** 6 **Inch** **Length**        LNTH 20 FEET 6 INCH
8 **Feet** 4 **Inch** **Height**        HGHT 8 FEET 4 INCH

Finally, solve for the number of bricks as follows:

**Conv** **Length** *(Blocks)*        BLKS 1171.429

which is rounded to: 1172 bricks.

Multiplying by the cost of a single brick using the calculator's "COST" function:

1 1 7 2 × 0 . 4 5 **Conv** 0 *(Cost)*        $527.$^{40}$

---

**Calculator Tip**

The Blocks function can also be used to determine needed quantities for counter or floor tiles, pavers, glass block, or any non-overlapping square or rectangular material. Simply set the Block Area to the size of the material desired.

**Calculator Tip:**

**Power User Tip:** 4065 v3.1 can solve for bricks based on values entered in **Length** only (for a single course), or **Length** and **Height** (for a wall area); keystrokes on older models differ. See your User's Guide for details.

## NOTES

**Calculator Version Note**

**4065 v3.0:** The list of results for `Height` is based on 4065 v3.1. Users of previous models should refer to `Height` key definition in User's Guide for more details.

**Calculator Tip**

**Weight per Volume:** The calculator's default setting for weight per volume is 1.5 tons per cubic yard.

**Calculator Tip**

**Power User Tip:** The calculator can be set to automatically display Square or Cubic solutions in the Units of your choice; see *Setting Preferences* in your User's Guide.

---

When the exterior of a residential or light commercial building is to be faced with brick, width of the foundation wall is generally increased so that the courses of brick sit directly atop the wall. This also requires an increase in footing width, the size of which is determined by the additional weight of concrete in the wall plus the weight of bricks and mortar. The volume-to-weight calculation is the same as that used in Chapter 4 for poured concrete and concrete block.

### ■ EXAMPLE 8.2

The weight density of brick w/mortar is approximately 120 lb per cubic foot. What is the total weight of the face wall in Example 8.1? Use the standard brick width of 4 inches.

**Solution:**

| KEYSTROKE | DISPLAY |
|---|---|

Enter the weight density using the keystroke sequence below. Continue to press the **0** key until the correct dimensional units (lb per cubic foot) appear on the display:

| 1 2 0 `Stor` 0 | `STORED` 120 LB PER CU FEET |
|---|---|

Next calculate the volume of the face wall in cubic feet by entering length, width and height of the wall:

| 2 0 `Feet` 6 `Inch` `Length` | LNTH 20 FEET 6 INCH |
|---|---|
| 4 `Inch` `Width` | WDTH 4 INCH |
| 8 `Feet` 4 `Inch` `Height` `Height` | VOL 2.109053 CU YD |
| `Conv` `Feet` | 56.94444 CU FEET |

Finally, convert this volume to pounds:

| `Conv` 4 *(lbs)* | 6833.333 LB |
|---|---|

---

Bricks are also used as pavers on horizontal surfaces such as hearths, walkways and patios. For this application, area coverage is generally taken as 32 square inches ( 8″ long × 4″ wide exposed face ) for a single brick.

### ■ EXAMPLE 8.3

A walkway that is 4′ wide and 60′ long is to be paved using standard bricks. Allowing a 5% waste allowance, find the number of bricks required for this project. If the bricks cost $0.52 each delivered on-site, find the total material cost for this walkway.

**Solution:**

KEYSTROKE                                                         DISPLAY

Enter the area coverage of a single brick:

③ ② **Inch** **Inch** **Stor** ④                 **STORED** B—AR  32. SQ INCH

Calculate the area using the given length and width. <u>Note that "width" should be entered using the **Height** key as if this were a wall</u>:

⑥ ⓪ **Feet** **Length**                                 LNTH 60 FEET 0 INCH
④ **Feet** **Height**                                      HGHT 4 FEET 0 INCH

Solve for the required number of bricks, and add a 5% waste allowance:

**Conv** **Length** (Blocks) ➕ ⑤ **%**                                                1134

Multiply by the unit cost using the "Cost" function:

✖ ⓪ • ⑤ ② **Conv** ⓪ (Cost)                                       $589.$^{68}$

Note that this method may be used for any type of paving or facing material simply by entering the appropriate coverage area of a single paver/face unit, rather than the 21 square inch or 32 square inch coverage for brick.

> **Calculator Tip**
>
> **Block Area:** The calculator's default setting for block/brick area is 128 square inches.

## PROBLEMS FOR CHAPTER 8

**P8.1.** An interior wall 12'-4" long and 7'-8" high is to be faced with bricks costing $0.48 each. Find the total number of bricks required and their cost.

**P8.2.** Find the extra weight carried by the floor for the wall of problem P8.1 if the weight density of the bricks w/mortar is 120 lb/cu. ft.

**P8.3.** A restaurant owner plans to construct two brick planters for use as room dividers. Each planter will be 10' long by 3' wide by 30" high. Including a 2% waste allowance, find the number of bricks needed for this project.

**P8.4.** The front of a light commercial building is 52'-6" long and 14'-7" high, and will be faced with bricks costing $0.42 each. Including a 3% waste allowance, calculate the number of bricks required, and their total cost.

**P8.5.** Using the same weight density as given in problem P8.2, find the weight of the face wall described in problem P8.4.

**P8.6.** A walkway 20′ long and 40″ wide will be built using paver bricks. How many pavers will be required?

**P8.7.** A patio that is 20′ long and 15′ wide is to be constructed using paver bricks costing $0.54 each. Find the number of pavers needed and their cost. Include a 2% waste allowance.

**P8.8.** The patio of problem P8.7 can also be constructed using specialty pavers that are 12″ long × 8″ wide and cost $1.25 each. For the same waste allowance, find the required number of pavers and their cost.

# CHAPTER 9

# ROOFING

Fiberglass *shingles* are the most common roofing material used today. These are not sold by the piece, but rather by their area *coverage*. One *square* of shingles will cover a 100 square foot area; a *bundle* of shingles covers an area of 33.33 square feet.

Your *Construction Master Pro* calculator will determine the number of 4′ × 8′ sheets of sheathing and the number of bundles or squares of shingles required to construct and cover a standard gable-end style roof. For cost estimating on new-house construction, only the floor dimensions and roof pitch are required, this information readily available from the architect's drawings. For existing homes on which the roof is to be replaced or recovered, it is more common to use the rise, run, and floor area covered by the roof to determine total material requirements.

### ■ EXAMPLE 9.1

A new ranch-style house 52′ long × 28′ wide will have a shingled roof with a 5″ pitch. If the shingles to be used cost $15.35 per bundle, find the number of bundles required and their total cost.

### Solution:

| KEYSTROKE | DISPLAY |
|---|---|

First, enter the pitch and floor dimensions for the house. (Because the pitch on both sides of the roof is assumed to be the same, we are able to use the total plan area here. For different pitches, one-half the roof area would be used in two separate roofing calculations.)

| | |
|---|---|
| 5 Inch Pitch | PTCH 5 INCH |
| 5 2 Feet Length | LNTH 52 FEET 0 INCH |
| 2 8 Feet Width | WDTH 28 FEET 0 INCH |

**Calculator Tip**

**Square of Shingles:** The calculator's default setting for one square of shingles is 100 square feet of coverage.

**Calculator Tip**

**Bundle of Shingles:** The calculator's default setting for a bundle of shingles is 33.33 square feet of coverage.

**Calculator Tip**

**Roof/Plan Area:** If the area is known, enter the value and then press **Conv** **Diag** (*Roof*).

Next, find the roof area, the number of roofing squares and bundles:

| KEYSTROKE | DISPLAY |
|---|---|
| Conv Diag (Roof) | ROOF 1577.333 SQ FEET |
| Diag | SQRS 15.77 |
| Diag | BNDL 47.32 |
| Diag | B—SZ 33.33 SQ FEET |
| Diag | 4×8  49.29 |
| Diag | STORED PTCH 5 INCH |
| Diag | PLAN 1456. SQ FEET |

Rounding the number of bundles to 48, multiply by the unit cost using the "Cost" function:

4 8 X 1 5 • 3 5 Conv 0 (Cost)    $736.$^{80}$

If the floor area is known, it may be entered directly without calculation.

### ■ EXAMPLE 9.2

A gable-end colonial home is to be built having a shingled roof with a 7-1/2" pitch. The plan area of this house is 1250 square feet. Calculate the number of 4 × 8 sheets and bundles of shingles required for construction of the roof.

**Solution:**

| KEYSTROKE | DISPLAY |
|---|---|

Enter the pitch and plan area, then press Diag to solve for the roof values:

| 7 Inch 1 / 2 Pitch | PTCH 7-1/2 INCH |
|---|---|
| 1 2 5 0 Feet Feet Conv Diag (Roof) | ROOF 1474.06 SQ FEET |
| Diag | SQRS 14.74 |
| Diag | BNDL 44.22 |
| Diag | B—SZ 33.33 SQ FEET |
| Diag | 4×8  46.06 |
| Diag | STORED PTCH 7-1/2 INCH |
| Diag | PLAN 1250. SQ FEET |

When the rise and run of a roof and the plan area or floor dimensions are known, calculations of roofing requirements for a particular structure are similar to those of the preceding examples.

### ■ EXAMPLE 9.3

The gable roof on a residential structure has a rise of 11'-6", a run of 14'-8", and covers a floor area of 1760 square feet. Rolled felt paper or plastic will be installed under the shingles, which will themselves be architectural-grade and cost $23.85 per bundle. Calculate the roof area, as well as the number of 4 × 8 sheets and number of bundles of shingles required to cover this roof. What is the total cost for shingles?

**Solution:**

| KEYSTROKE | DISPLAY |
|---|---|

Enter the rise, run, and floor area:

| 1 1 Feet 6 Inch Rise | RISE 11 FEET 6 INCH |
|---|---|
| 1 4 Feet 8 Inch Run | RUN 14 FEET 8 INCH |
| 1 7 6 0 Feet Feet Conv Diag (Roof) | ROOF 2236.515 SQ FEET |
| Diag | SQRS 22.37 |
| Diag | BNDL 67.10 |
| Diag | B—SZ 33.33 SQ FEET |
| Diag | 4×8  69.89 |
| Diag | STORED PTCH 9-7/16 INCH |
| Diag | PLAN 1760. SQ FEET |

Rounding the number of bundles to 68, and multiplying by the unit price using the calculator's "Cost" function:

| 6 8 × 2 3 • 8 5 Conv 0 (Cost) | $1621.⁸⁰ |
|---|---|

## PROBLEMS FOR CHAPTER 9

**P9.1.** A 26' wide × 42' long house has a 4-1/2" pitch gable roof. Calculate the number of 4 × 8 sheets and the number of squares of shingles required for this roof. What is the roof area?

**P9.2.** The plan area for a house having a 6" pitch gable roof is 1400 square feet. Unit cost of the shingles to be used is $16.25 per bundle. Find the number bundles and 4 × 8 sheets required for this roof, as well as the total cost of shingles.

**P9.3.** A gable roof with a 6′-4″ rise and 15′-3″ run covers a floor area that is 30′-6″ wide × 54′-8″ long. Architectural-grade shingles at $22.75 per bundle will be installed over a felt paper base. Determine the number of 4 × 8 sheets and square footage (roof area) of felt paper required. How many bundles of shingles are needed, and what is their total cost?

**P9.4.** A floor area of 2250 square feet will be covered by a gable roof having a rise of 8′-3″ and a run of 17′-6″. How many 4 × 8 sheets and squares of shingles are needed for this roof?

**P9.5.** The gable roof covering an 80′ long × 32′ wide horse stable has a pitch of 30°. The roof will consist of 4 × 8 sheets of plywood covered by 4 × 8 sheets of corrugated steel roofing. What is the roof area, and how many 4 × 8 sheets of each material will be required?

**P9.6.** A gable roof with 38° pitch covers a plan area of 1850 square feet. Determine the roof area, number of squares of shingles, and number of 4 × 8 sheets required.

# CHAPTER 10

# DRYWALL

*Drywall* is the most common wall material used today, and is available in several sheet sizes, including 4′ × 8′, 4′ × 9′, and 4′ × 12′. These sheets may be purchased in a variety of thicknesses, with one-half inch finding the widest use.

For a given wall area, your *Construction Master Pro* calculator can determine the number of drywall sheets required in each of the three sizes listed above. Area may be entered (a) as an already computed value in total square feet, or (b) in the form of room dimensions with area yet to be computed. Once the area has been entered, however, pressing the **Conv** **Height** keys initiates calculation of the required number of drywall sheets for a particular size. Since the order in which sheet sizes are displayed is based on the last displayed value from previous calculations, successively pressing the **Height** key will yield the required number of sheets for the remaining sizes.

### ■ EXAMPLE 10.1

A room that is 18′-4″ long × 14′-6″ wide has walls that are 9′-0″ high. Determine the number of drywall sheets needed to cover these walls in each of the three most common sheet sizes.

**Solution:**

| KEYSTROKE | DISPLAY |
|---|---|

First, enter the room dimensions and compute the room perimeter:

| 1 8 Feet 4 Inch Length | LNTH 18 FEET 4 INCH |
| 1 4 Feet 6 Inch Width | WDTH 14 FEET 6 INCH |
| Width Width Width | PER 65 FEET 8 INCH |

Next, enter the wall height, and determine the total wall area:

| 9 Feet Height | HGHT 9 FEET 0 INCH |
| Height Height | WALL 591. SQ FEET |

> **Calculator Version Note**
>
> **4065 v3.0:** The list of results for **Height** is based on 4065 v3.1. Users of previous models should refer to **Height** key definition in User's Guide for more details.

> **Calculator Tip**
>
> **Drywall:** After displaying all three drywall sizes, an additional press of [Height] displays the entry used for the calculation.

Finally, solve for the number of sheets in each size:

| KEYSTROKE | DISPLAY |
|---|---|
| [Conv] [Height] (Drywall) | 4×8  18.46875* |
| [Height] | 4×9  16.41667* |
| [Height] | 4×12  12.3125* |

(* Note that these numbers would be rounded to 19, 17, and 13 respectively.)

Drywall is also used as a ceiling material. Due to the difficulty of installing these relatively heavy sheets overhead, 4′ × 8′ is the most popular size for this application.

■ **EXAMPLE 10.2**

The ceiling of the room in Example 10.1 is to be covered with one-half inch thick drywall that costs $9.60 per 4 × 8 sheet. Find the number of sheets needed, and the total material cost.

**Solution:**

| KEYSTROKE | DISPLAY |
|---|---|

Enter the room dimensions and determine the total ceiling area:

[1] [8] [Feet] [4] [Inch] [Length]     LNTH 18 FEET 4 INCH
[1] [4] [Feet] [6] [Inch] [Width] [Width]     AREA 265.8333 SQ FEET

Find the required number of sheets:

[Conv] [Height] (Drywall)     4×8  8.30729

Rounding the number of sheets to 9, calculate the total cost:

[9] [×] [9] [.] [6] [0] [Conv] [0] (Cost)     $86.40

For wall heights that are equal to, or slightly less than, the 8′, 9′, or 12′ sheet lengths, the number of sheets required may also be determined using linear dimensions rather than areas. Because they are not based on area, the number of sheets displayed for each size will be the same.

■ **EXAMPLE 10.3**

A hotel conference room is to be 49′-10″ long × 31′-8″ wide and will have walls that are 11′-9″ high. How many 4 × 12 sheets of drywall will be required for the walls in this room?

**Solution**

Calculate the perimeter of this room:

| KEYSTROKE | DISPLAY |
|---|---|
| 4 9 Feet 1 0 Inch Length | LNTH 49 FEET 10 INCH |
| 3 1 Feet 8 Inch Width | WDTH 31 FEET 8 INCH |
| Width Width Width | PER 163 FEET 0 INCH |

Now, solve for the number of sheets:

| | |
|---|---|
| Conv Height *(Drywall)* | 4×8  40.75 |
| Height | 4×9  40.75 |
| Height | 4×12  40.75 |

## PROBLEMS FOR CHAPTER 10

**P10.1.** A wall has an area of 185 square feet. How many sheets of drywall in each of the three popular sizes is required to cover this wall?

**P10.2.** For a wall whose dimensions are 16′-9″ by 8′-6″, find the number of drywall sheets in 4 × 8, 4 × 9, and 4 × 12 sizes that are needed to cover the wall.

**P10.3.** A room that is 15′-11″ long × 12′-7″ wide has walls that are 9′-0″ high. Calculate the wall area, and the number of 4 × 8 sheets necessary to cover these walls.

**P10.4.** Find the ceiling area of the room in problem P10.3. If half-inch thick drywall is used in 4 × 8 sheets at a cost of $9.60 per sheet, determine the number of sheets required and their cost.

**P10.5.** A room is 22′-3″ long, 17′-5″ wide, and has walls that are 10′-0″ high. Find the total wall area, and the ceiling area. If 5/8″ thick 4 × 9 sheets of drywall costing $13.50 per sheet are used to cover all surfaces, calculate the total number of sheets needed and their cost.

**P10.6.** For the room in problem P10.3, find the required number of 4 × 9 sheets based on the linear dimensions of the walls.

**Correction Sheet**
Issued 25 August 2006

# ERRATA SHEET FOR CONSTRUCTION MASTER® PRO WORKBOOK AND STUDY GUIDE

## Corrections to "Glossary of Construction Terms"

**Heel**  The definition shown for Heel is actually the Seat Cut. The Seat Cut is the cut that a rafter rests on when sitting on the top plate of a wall.

**Heel Cut**  Definition should be corrected to say, "The vertical cut at the end of the rafter that forms the bird's mouth with the **Seat Cut**.

## Corrections to Chapter Examples

**Example 7.1**  The example is missing a keystroke during the calculation of Stairs. Directly after display of the Stringer Length, the Inclination Angle is displayed (answer: **INCL 36.18°**), and then the Run is displayed.

## Corrections To Chapter Problems

**P1.7**  The answer 22,029.23 SQ FT should read **29,029.23 SQ FT**. Also, the diagonal measurement for segment G - F on *Figure 1-11* should read 82.20'.

**P6.6**  The pitch answer refers to the computed <u>Hip Rafter Pitch</u> and not the Inch Pitch as displayed on the calculator. 8-1/2" is the correct answer for the Hip Rafter Pitch as it relates to the 17" hip unit run. After calculating your Hip/Valley results, you can solve the Hip Rafter Pitch by entering the calculated Hip/V Plumb Angle as the Slope (i.e. `Conv` `Pitch`), then solve for Hip Rafter Pitch (in inches) with repeated presses of the `Pitch` key.

**P7.2**  The answer for Stairwell Opening should read **12' 3"** (not 12' 1").

**P10.3**  The answer for 4 x 8 Sheets should read **17**. (The answer is 16.03125 and then you round to 17 sheets.)

**P10.4**  The value of the cost was omitted. The cost is **$67.20**.

**General Note on Rounding:** If you re-enter an intermediate answer or an answer mid-way through an example and use it to compute other solutions, your answer MAY differ in the 4 or 5th decimal place. This is due to internal calculator rounding and is not significant.

For the latest updates and other helpful tips, please go to our website at: **http://www.calculated.com/CMProWorkbook**

# ANSWERS TO CHAPTER PROBLEMS

## Chapter 1

P1.1  247.50 FT

P1.1  286.875 FT

P1.3  2883.982 SQ FT = 320.4424 SQ YD

P1.4  17,856 SQ FT = 1984 SQ YD

P1.5  128,250 SQ FT

P1.6  Area = 6824.98 SQ FT; Perimeter = 452.3485 FT

P1.7  22,029.23 SQ FT

P1.8  31,666.91 SQ FT

P1.9  3152.039 SQ FT

P1.10 approximately 32,400 SQ FT

## Chapter 2

P2.1  59,375 CU YD

P2.2  3076.774 CU YD

P2.3  1386.667 CU YD

P2.4  80.75148 Ton

P2.5  (a) 28°27′36″

(b) 34.21°

P2.6  RISE 40 FT 9-5/8 IN

P2.7  %GRD 24.78386; PTCH 13.92°

P2.8  RUN 40 FT 0 IN; 24.69136 CU YD

P2.9  Existing: 23.31797%; Proposed: 34.37788%; 4822.222 CU YD

## Chapter 3

P3.1  17.18529 CU YD

P3.2  (a) 16.50752 CU YD

(b) 15.63947 CU YD

(c) 5.550%

P3.3  0.327249 CU YD

P3.4  (a) 0.843576 CU YD

(b) 362.0387 SQ FT

P3.5  18.53873 CU YD

P3.6  139.6263 CU YD

P3.7  (a) 5.812012 CU YD

(b) 342.3803 SQ FT

## Chapter 4

P4.1  (a) 25.92593 CU YD

(b) 3.45679 CU YD

(c) 11.7647%

P4.2  119000.$^{02}$ LB

P4.3  (a) 3.703704 CU YD

(b) 2.222222 CU YD

(c) $562.96

P4.4  24,800 lb

P4.5  (a) 153 Blks

(b) $214.20

(c) $211.11

(d) 24.45112% less expensive

P4.6  16,100 LB

## Chapter 5

P5.1  (a) 16 Studs

(b) 14 Studs

(c) 18 Studs

P5.2  (a) 5.333333 BD FT

(b) 12 BD FT

(c) 13.33333 BD FT

(d) 23.33333 BD FT

P5.3  (a) $70.00

(b) $413.28

(d) $344.96

P5.4  423.5 BD FT; $143.99

P5.5  434 BD FT; $180.11

P5.6  1288 BD FT; $547.40

## Chapter 6

P6.1  Rise = 6′-1 5/16″; rafter length = 14′-8″; plumb = 24.62°; level = 65.38°

P6.2  Pitch = 6-5/8″ or 28.89°; rafter length = 17′-5″; plumb = 28.89°; level = 61.11°

# Answers to Chapter Problems

**P6.3** Run = 11'-0"; rafter length = 11'-7 1/8"; plumb = 18.43°; level = 71.57°

**P6.4** Common rafters:
pitch = 26.57°;
length = 17'-7 5/16"

plumb = 26.57°;
level = 63.43°

Hip/valley rafters:
length = 23'-7 1/2";
plumb = 19.47°

level = 70.53°;
cheek = 45.00°

**P6.5** Jack rafters @ 16" o.c.:
16'-1 7/16"; 14'-7 9/16"; 13'-1 5/8"; 11'-7 3/4"; 10'-1 7/8"; 8'-8"; 7'-2 1/16"; 5'-8 3/16"; 4'-2 5/16"; 2'-8 7/16"; 1'-2 9/16"

Jack rafters @ 24" o.c.:
15'-4 1/2"; 13'-1 5/8"; 10'-10 13/16"; 8'-8"; 6'-5 1/8"; 4'-2 5/16"; 1'-11 1/2"

**P6.6** Hip/valley rafters:
pitch = 8-1/2";
length = 13'-10 1/4";

plumb = 35.26°;
level = 54.74°

**P6.7** Jack rafter lengths:
8'-5 13/16"; 5'-7 7/8"; 2'-9 15/16"

**P6.8** Irregular hip/valley rafters:
length = 15'-6";
plumb = 18.28°;

level = 71.72°;
cheek 1 = 37.57°;
cheek 2 = 52.43°

Irregular jack rafters:
8'-5 1/2"; 6'-8 1/2"; 4'-11 1/2"; 3'-2 1/2"; 1'-5 1/2";
plumb = 28.44°;
level = 61.56°;
cheek = 37.57°

Regular jack rafters:
9'-9 7/8"; 7'-0 1/16"; 4'-2 1/4"; 1'-4 7/16";
plumb = 22.62°;
level = 67.38°;
cheek = 52.43°

**P6.9** Rake-wall stud lengths:
3'-10 5/16"; 3'-2 9/16"; 2'-6 7/8"; 1'-11 1/8"; 1'-3 7/16"; 0'-7 11/16";
cheek = 17.82°

**P6.10** Rake-wall stud lengths:
12'-2 3/4"; 11'-7 7/16"; 11'-0 1/16"; 10'-4 3/4"; 9'-9 7/16"; 9'-2 1/16"; 8'-6 3/4";
cheek = 24.62°

## Chapter 7

**P7.1** Risers: 15 @ 7-7/16" height, underage of 7/16"

Treads: 14 @ 10" width, no overage/underage

Stairwell opening: 10'-1"; Stringer length: 14'-6 1/2"

Inclination angle: 36.64°; Run: 11'-8"

**P7.2** Risers: 16 @ 7" height, no overage/underage

Treads: 15 @ 10-1/2" width, no overage/underage

Stairwell opening: 12'-1"; Stringer length: 15'-9 5/16"

Inclination angle: 33.69°; Run: 13'-1 1/2"

**P7.3** Risers: 20 @ 7-1/2" height, no overage/underage

Treads: 19 @ 10-1/8" width, overage of 3/8"

Stairwell opening: 10'-1 1/2"; Stringer length: 19'-11 3/8"

Inclination angle: 36.53°; Rise: 12'-6"

**P7.4** Risers: 12 @ 7-1/2" height, no overage/underage

Treads: 11 @ 11-3/4" width, overage of 1/4"

Stairwell opening: 11'-9"; Stringer length: 12'-9 5/16"

Inclination angle: 32.55°

**P7.5** Risers: 14 @ 8" height, no overage/underage

Treads: 13 @ 9" width, no overage/underage

Stairwell opening: 8'-5 1/4"; Stringer length: 13'-0 9/16"

Inclination angle: 41.63°; Run: 9'-9"

**P7.6** Risers: 12 @ 7-1/2" height, no overage/underage

Treads: 11 @ 11" width, no overage/underage

Stringer length:
12'-2 7/16"; Inclination angle: 34.29°

Run: 10'-1"

## Chapter 8

**P8.1** 649 bricks @ $311.52

**P8.2** 3782.222 LB

**P8.3** 910 bricks

**P8.4** 5408 bricks @ $2271.36

**P8.5** 30,625 LB

**P8.6** 300 pavers

**P8.7** 1377 pavers @ $743.58

**P8.8** 459 specialty pavers @ $573.75

## Chapter 9

**P9.1** 4 × 8 sheets: 36.45; 11.66 squares; area: 1166.257 square feet

**P9.2** 47 bundles; 4 × 8 sheets: 48.91; $763.75

**P9.3** 4 × 8 sheets: 56.42; roof area: 1805.403 square feet; 55 bundles; $1251.25

**P9.4** 4 × 8 sheets: 77.73; 24.87 squares

**P9.5** roof area: 2956.033 square feet; 4 × 8 sheets: 92.38

**P9.6** roof area: 2347.684 square feet; 23.48 squares; 4 × 8 sheets: 73.37

## Chapter 10

**P10.1** 4 × 8: 6 sheets; 4 × 9: 6 sheets; 4 × 12: 4 sheets

**P10.2** 4 × 8: 5 sheets; 4 × 9: 4 sheets; 4 × 12: 3 sheets

**P10.3** Wall area: 513 square feet; 4 × 8 sheets: 7

**P10.4** Ceiling area: 200.2847 square feet; 4 × 8 sheets: 7

**P10.5** Wall area: 793.3333 square feet; ceiling area: 387.5208 square feet; 4 × 9 sheets: 33; $445.50

**P10.6** 4 × 9 sheets: 15